One Cap Over The Windmill

One Cap over the Windmill

Kenneth Hutchison

Whittles Publishing

Typeset by
Whittles Publishing Services

Published by
Whittles Publishing,
Roseleigh House,
Latheronwheel,
Caithness, KW5 6DW,
Scotland, UK

ISBN 1-870325-52-4

Printed by J.W. Arrowsmith Ltd., Bristol

For Caitlin and Michael,
Louise and Robbie

(Photograph courtesy Paul Graham)

A short explanation…

The saying to *throw one's cap over the windmill* means to act recklessly and impulsively and it provides an apt title for the book, reflecting the author's impetuous pursuit of the property which he secured and restored, seemingly against the odds.

Contents

Acknowledgements

I am indebted to Elizabeth Adair, Arlene, Mum and Dad, Sylvia and Ian, Jacqueline and Gerard (the Fitzies), BIll Davidson, Kay Hogg Edna South and John Hardy. To anyone who goes uncredited for their contribution to this book, I apologise. Special thanks to the publishers, without whom I may never have told the story.

1 Football

It was a freezing cold morning in February 1973 and the match had been given the go-ahead after a brief pitch inspection, which clearly had not taken into account the ice-hardened ground or the uphill camber towards the north goal-mouth, but the cubs were tough as old boots and quite used to playing football on more treacherous surfaces than Edgehead Park. I knew that our opponents, Edgehead, had struggled to get a team together because they were lucky if they had eleven children in the village, and even if they did there was a good chance that most of them would be uncoordinated, underage and underdeveloped, probably ill-equipped with unfashionable sportswear and straw sticking out of their ears. The Edgehead kids were so unfashionable, as I recall, that they didn't even wear flared trousers to school. We had a horribly prejudiced image of children from the countryside, and usually we weren't far wrong.

The Dalkeith St Johns & Kings Park 12th Midlothian Cub Football Team, on the other hand, was a whole different ball game. For as long as I had been in the team we had won every single match we played, every tournament and every league. Our scores were outrageous. It was not uncommon for us to be celebrating a win of 10-nil or 13-2. We were the absolute champions and I was sure even then, that some of our players would one day go on to great things in the football world. And they did. Peter Docherty played for the Hibs reserves in his late teens, Gogsie Flynn and Mark Gilhooly became formidable regulars in their school team at St David's High, and John Landells spent ten years at Bonnyrigg Rose, accumulating a mantelpiece full of glittering trophies.

As in nearly every sporting success story, the key to our success was our coach, Inglis Hutchison. He was thirty-six, handsome, friendly and enthusiastic, and had black curly hair with fashionable red bushy sideburns. He drove a D registration navy blue Ford Anglia and always supplied juice and quartered oranges for the boys at training and at matches. He was on the committee of Dalkeith Thistle FC so he knew all the tactics and strategies that real teams used to win games. Our own weekly training sessions at Ironmills Park were rigorous and relentless as I remember, and when injury struck, Inglis would bring out the 'magic sponge' to cure any football casualty in seconds. He would allow seriously hurt players, who were crying in agony sometimes, to say 'oh ya bugger' and promised not to tell their parents. This would turn their tears to laughter and alleviate the pain, and within seconds the wounded soldiers would be chasing the ball round the park again in swarms of twenty.

I usually played right back when I wasn't the substitute, but I was still proud to feel part of the team. What made me most proud though, was that Inglis Hutchison was my Dad, and if the truth be known, that's the only reason I ever made it into the team. Yes, I had two left feet, a distinct disadvantage for a right back, and I knew that if it wasn't for downright nepotism, I'd never get picked. Even the desperate Edgehead would have left me on the bench. Dad would give me every opportunity to excel or at least to improve at the game but we both knew I couldn't concentrate for more than a minute or two. For I was too open to distraction and my imagination would run wild, and I found myself following the adventures of a pheasant or a rabbit, or staring upwards at a light aeroplane circling miles above us, or watching a girl on a swing, or even the Sunblest van coming up the village.

Regularly, during a match, I would wake from my daydreams to the deafening sound of my name being yelled from the sidelines by my Dad while the rest of the team were clocking up the goals. If ever a goal *was* scored against us it was usually my fault or I had scored it. I was the only cub on the park to go ninety minutes with clean knees every week, and clean boots for that matter, because no-one was stupid enough to pass the ball to me. So every Saturday, while the game went on around me, I would lose myself in the pleasure of being in another town or village, and take in the atmosphere.

The 12th Midlothian Cub Team. The author sports the black tracksuit.

I remember Edgehead being a long journey in a Ford Anglia packed with cubs, hardly able to climb the hill, in second gear all the way, and travelling at the speed of a funicular railway. From Dalkeith it was *away* out in the country and when we got there the village seemed so rustic, like something from a story book. The trees were the tallest I had ever seen, the whole forest reaching all the way up to the cloudless blue sky. There were spectacular views of fields and hills in the distance and all the houses were little cottages, all different shapes and colours, no two alike. The birds sang different tunes and there were animal noises I did not recognise. It was like being in another country. I was in Pogles Wood. Or Munchkinland. But the big distraction that really sealed my fate as a non-contender for 'man of the match' against Edgehead that day was the sight of an unusual building on the edge of the wood surrounded by a wooden fortress fence with a big white gate. I could see it clearly from the eighteen yard box; it was in the shape of a giant helter-skelter and made of irregular sandstone boulders with little old-fashioned windows that formed eyes and a mouth and smiled over the park like a cartoon house. The huge face wore a pointed cone-shaped hat of grey slate and its tall chimney was the feather in its cap. Adjoining the stony tower was a sprawling white castle-like wing, trimmed with black wrought iron stars and thistles, and balconies looking south over the hills and west into the woods. Would I catch a glimpse of Oberon and Titania looking out over the countryside? Had the old woman who lived in a shoe flitted? Images of Hansel and Gretel and evil woodcutters flashed through my mind as I half-heartedly chased after a ball that accidentally came into my territory from high above. I closed my eyes and booted it as a hard as I could. I had no control whatsoever over the direction it would take and when I opened my eyes again it was falling from the sky, yards out of play, before landing in the garden of the mysterious house. My jeering team-mates sent me off to retrieve the ball and for a second or two I wondered whether I would come back alive. I crept in through the gate – no sign of a witch – and I scurried towards the ball which had lodged itself in a bush. I picked it up and threw it (the way girls throw, apparently) over the fence and the game restarted immediately without me as I ran for my life back to my crucial position in defence. On my return, one of the Edgehead players accidentally passed the ball to me and I was about to volley it up the park with spectacular style and finesse when Justin Chater skilfully dribbled it away from me and meandered his way round all of our opponents who fell to the ground one by one like skittles, as he shot at the Edgehead goalmouth. The goalkeeper dived out of the way as the ball hit the target and rolled off under the floppy net, bouncing into the undergrowth in the woods behind. I convinced myself that I had played an important part in setting up this goal, then I turned my attention back to the strange house again for a while before the whistle blew three times. Someone told me the final score was 15-nil to us.

We had 'maulicated' them (a fashionable word in 1973 which, despite its graphic expressiveness, never quite made it from the ghettos of Dalkeith into the dictionary). We had cause for celebration, and as a bonus I was the only one who didn't have to nurse any cuts, bruises or scratches. A double triumph! As we began our descent towards Dalkeith, the pleasant images of Edgehead slowly evaporated. Already we were looking forward to next week's away match against Musselburgh.

2 House hunting

Twenty years later I was living in Aberdeen having studied music at college for three years, taught music at Bankhead Academy for three years and then escaped from teaching to become a disc jockey on Northsound Radio, the north east's number one commercial radio station. Number one because it was the *only* commercial radio station in the north-east of Scotland. It couldn't fail; it had the monopoly; there was *no* competition whatsoever and everybody listened. But the novelty wore off being a DJ when presenters' choice was abolished, and a new computerised system called 'Selector' determined the musical content of my daily programmes. Suddenly, my shows were far too unimaginative for my liking. Good records were banned. The Jam and The Specials and Squeeze were old hat, XTC and Joe Jackson were no-go areas and an influx of Stock, Aitken and Waterman prevailed. Time for me to go. Three years had been long enough anyway, and I returned to teaching, this time in the American School in Aberdeen (an educational establishment which served the precocious children of the oil industry barons, and those of the American Air Force based at Edzell near Montrose) for a further thirty-six months. On the positive side, I had succeeded in stretching my concentration span from thirty seconds as an eight-year-old footballer to a somewhat more respectable three year cycle before reaching my boredom threshold, and in 1993 it was time for yet another change. Only this time my desire was to return to Edinburgh.

To this day I can never actually remember leaving home and making the initial move to Aberdeen. I simply went to college there and woke up some twelve years later to find that time had marched on. Suddenly I was old (mid-twenties) and I had a wife, two cats and a mortgage. Arlene, my wife, was working as a continuity announcer for Grampian Television, linking the programmes and reading the news. She was well-respected as a television presenter and very popular in the north-east of Scotland but for her too it was time for a change. Away from the haggis and shortbread image, away from the 'Country Matters Special' featuring the natural habitat of the Caledonian wart-hog.

At the end of 1992 the offer came to Arlene of a job with Scottish Television, and this was the excuse we needed to make a move. I urged Arlene to take the job. I thought a teaching post for me would come up in the Edinburgh area eventually, and in the meantime we could sell our flat in Aberdeen and look for a new place to live at our leisure.

A few days before Christmas we were heading for my Mum and Dad's house in Woodburn, Dalkeith. When we reached Edinburgh I parked the big red Citroen CX in the middle of George Street, a feat in itself, and Arlene went off to do some last minute Christmas shopping in Jenners. With the real prospect of a house-move on my mind, I was about to peruse the property section of the *Evening News* while listening to Jay Crawford blethering on Radio Forth in the car. As I wrestled with the broadsheets, searching for pages of appropriate property, my eyes fell on one particular picture, no bigger or smaller than any of the others, but instantly more striking because of the house's shape, its charm, its appeal. It was a photograph of the old house I had seen so many years before in Edgehead.

I was immediately transported back to 1973, a wee boy staring up at the imposing structure of this enchanting house, lost in daydreams of giants and hobgoblins and ghosts and fairies. And now, exactly twenty years later, it was on the market. Every other picture on the property page disappeared and I found out that this place was called 'The Old Windmill'. What a great name. I realised for the first time that it had been a working windmill in the distant past, with massive rotating sails, the mechanism, the lot. The information in the advertisement was sketchy, but at least I knew exactly how to get to the site of the windmill. I could drive straight there if I wanted to. Best wait for Arlene, though, I thought.

At the bottom of the advert was a contact number for Ryden and Company, the agents who were selling the property. My heart was thumping because I had a hunch that already I had found our house 'in Edinburgh', and although I knew that the idea of living there was pie in the sky, I was interested enough to at least think about making an inquisitive phone call to establish how realistic it would be for us to obtain such a dream house. Looking at my watch, I saw that it was a minute to five and very likely Ryden and Company would be closed in a matter of seconds. I cleared the steamed-up windows with the sleeve of my coat and peered out into the dismal street in search of a phone box on the off chance that I could get some information there and then. There wasn't a public telephone in sight but, miraculously, directly in front of me, was the Edinburgh office of Ryden. In disbelief I looked down at the newspaper, up at the shop again and then I glanced over at the digital clock on the dashboard. It read exactly five o'clock and I could see Ryden's fluorescent striplights flickering off before my eyes. I threw the *Evening News* up in the air, pages flying in and out of the car as I kicked open the door and sprinted across George Street in the wind and rain, narrowly missing being killed by a maroon bus.

The shop door was already locked. There was no-one around. I couldn't believe that anyone could leave a building so quickly, so I battered on the glass door with the palms of my hands. After thirty seconds or so I was about to run away to find the rear entrance to the shop (probably in a lane up Queen Street I guessed) to intercept the invisible shop assistant, when from out of the darkness there appeared a middle-aged woman who was in the process of putting on her coat while rummaging in her bag for keys.

Somehow I managed to convince the lady that I was not a mugger or an armed robber and requested a schedule for The Old Windmill as advertised in the *Evening News*. Beyond the call of duty, she invited me into the shop and fingered her way through a four-storey filing cabinet. It was apparent she could not find what she was looking for. It was that sinking feeling for me and my heartbeat slowed down to a normal pace. Her theory was that the schedules had been removed when the house was sold. There were none on display in the window and none in the cabinet. She even searched her stationery drawer, just in case. I knew it had been too good to be true. The prospect of acquiring our dream house had come and gone in less time than it had taken Arlene to buy a pair of socks for her brother's Christmas present.

I thanked the lady for her time, and on parting she suggested that I phone the office in the morning, by which time she *might* have obtained an explanation of why The Old Windmill was advertised for sale in the newspaper but was no longer on file. Off she ran to St Andrew's Square for her bus and I returned to the car. In my blind panic I had left the car door open and the *Evening News* had blown away all over Edinburgh except, strangely enough, the page with The Old Windmill on it. This was quite convenient because when Arlene returned I was at least able to show her the picture and explain the last few minutes of my life, and how it might be a good idea to follow up the sale of the windmill. Arlene clunked the door shut and clicked in the seat belt and studied the advert in the paper, saying she liked the look of the house, but she didn't get over-excited. I confirmed that I would phone Ryden in the morning anyway, and commented on how nice Stuart's socks were.

The big Citroen pulled away from the middle of George Street and lumbered towards the A68 and Dalkeith. On the way home, Arlene read to me from the soggy remaining page of my *Evening News*. I pretended to show an interest in a semi-detached bungalow in Duddingston.

...this place was called 'The Old Windmill'. What a great name.

PROPERTY PARTICULARS

Ryden
RESIDENTIAL
Incorporating LISTER MARSHALL

CHARTERED SURVEYORS

3 Big Ideas

The next day was a Saturday, the last before Christmas Day, and I knew that Ryden would be open for business. Having spent the night at Mum and Dad's house, I found myself in Dalkeith town centre early in the morning, posting my Christmas cards late as usual. The other incentive for me to get up early was of course that I knew I had to phone Ryden to solve The Old Windmill mystery. I stood in 'Jarnac Court', an exotic name for a shopping precinct in the middle of Dalkeith, the French connection having been made in the early 1960s when Jarnac and Dalkeith were twinned in an act of community spirit. However, Jarnac Court was not nearly as exotic as the French name would suggest. The council, in their wisdom, had flattened most of Old Dalkeith some years before, and in place of its charming historic market-town buildings they allowed the construction of the most tasteless angular architecture imaginable. But I have an elephantine memory and I can clearly remember the sights, sounds and smells of the town as it was even in the swinging sixties, virtually shop by shop, street by street, having spent the first seven years of my life toddling up and down the High Street, in and out of the ancient closes. There is no comparison between old and new, sadly.

There I stood in a phone box opposite the bookmakers (which I fondly remembered had been a motorbike shop when I was at school) and the cubicle smelled of urine from the night before. The phone boxes in Dalkeith now seem to double as public conveniences at weekends. From my pocket I pulled the remaining page of last night's *Evening News*, which was by now dog-eared origami, but Ryden's faded phone number was still readable. Having thumped the digits with my left hand, I put the phone to my right ear and stuffed my left index finger in my left ear in an attempt to drown out the noise of the nose to tail traffic roaring its way through the town centre. Expecting a short negative response and a quick call I just dropped a ten pence coin in the slot. The phone was answered by the anonymous helpful woman I'd met the previous day; only this time she had better news for me, which took me by surprise. She told me that there had been so much interest in The Old Windmill in the past few months that the supply of schedules had been exhausted. However, she thought that one copy might remain in the Ryden branch which was closest to the property itself. By an amazing coincidence this was the Dalkeith branch.

As she uttered the words 'Dalkeith branch' I looked up through a gigantic saliva gob on the plastic pane on the phone box to see, not more that ten yards away from me, the very Dalkeith branch of Ryden, right here in Jarnac Court! We even said 'Jarnac Court' together, in unison. When I had left for college, years earlier, that shop had been Charlie Watt's, an old-fashioned barber's shop (namesake of one Rolling Stones drummer) where the barbers wore maroon nylon jackets with black lapels, and Brylcreem. A fashionable haircut you could not expect but 'something for the weekend', a trim or a shave, or even a 'trend' could be bought here. It was worth paying the money to step back in time and sit in a proper barber's chair in a shop that had a proper red and white pole outside. I heard that Charlie Watt eventually died of a heart attack, and for a spell it was a highly unfashionable 'Boutique' (the Jarnac connection again?) before Ryden and Company moved in. I ran from the phone box in four big strides and on display in the window under the 'property for sale' section there it was – a full colour photograph of The Old Windmill accompanied by written details of the house and garden.

As I made my entrance I could tell that the typist didn't get many strangers round these parts. She stopped typing the way a bar-room pianist would have stopped playing the piano, and she looked up at me slowly as if I had just walked into a saloon in Deadrock. A deafening silence fell across the room. I thought about demanding a whisky in a dirty glass, but instead I asked if she had any information about the house on the hill. She turned out to be pleasant and helpful, and soon my wild west heroine was to be seen thumbing through her filing cabinet, from which she produced a beautiful glossy schedule of 'particulars of The Old Windmill, Edgehead'. It read:

> *For Sale. A unique country property. Restored and extended period property of enormous charm. Edinburgh 10 miles. Dalkeith 4 miles. City by-pass 10 minutes. This unique listed property was formed from a former tower mill in 1961* (year of completion) *and has subsequently been much written about and admired. It is set in a tranquil open glade amidst fine mature woodland. The property is situated off a private unadopted lane a short distance from the street that forms the little community of Edgehead.*
>
> *Edgehead lies approximately 10 miles south of Edinburgh amidst rolling well-wooded countryside. The village of Pathhead, within two miles, provides local shops and other amenities, whilst a wide range of facilities is available in Dalkeith. The nearby A68 is a quick route to the south, and the north leads via Dalkeith to Edinburgh City By-Pass and thence to the Central Scotlands motorway network. The surrounding countryside is some of the most attractive to be found within easy reach of Edinburgh and is noted for a number of excellent pheasant shoots. To the east lie many famous golf courses along the Firth of Forth coastline.*
>
> *"The Old Windmill" was originally created in 1961 by Elizabeth Adair, Scotland's first lady announcer of the BBC and an expert and devotee of gardens, animals and historic buildings. The white rendered brick cavity wall additions are carefully in keeping with one of Scotland's few remaining eighteenth century windmills and the property has had significant press coverage over the years. The full story of this restoration can be found in the book I Bought A Windmill which was written by Elizabeth Adair herself. Great care was also taken with the garden by the subsequent owners, which now boasts a wonderful selection of flowering shrubs and small trees,*

bounded by fine woodland. Double glazing is fitted, and there is a burglar alarm. Viewing strictly by appointment through the Agents above.

I took the last schedule in Deadrock and headed for the door. The woman's parting shot, 'It needs a lot of work and it's not everyone's cup of tea', fell on deaf ears.

The book. I had to get my hands on the book mentioned in the schedule, because I was sure this would give me all the information I needed to satisfy my curiosity about the house; information that estate agents, solicitors and even prospective buyers didn't know. It would also, if necessary, apply the brakes to my potentially dangerous imagination and lend a more practical and sensible thread to my investigations. Knowing that The Old Windmill was still on the market, I believed we were in the race, even if the odds on buying it were against us, and the thing to do now was to set the wheels in motion.

I was now in the Olympic Games of house buying, the obstacle race of property speculation and of course the egg and spoon race of flat selling in Aberdeen. Who were my competitors? How many people were interested in The Old Windmill? How interested were they? How could I eliminate them? A bizarre Christmas tree lights accident perhaps? And one small point. The asking price. Somehow this was not to be true. No, no, no. It had to change. The grim reality was that offers over 'an astronomical amount' were invited. Yet due to our financial situation only 'offers under' this figure could ever be submitted by us, and 'offers under' were unheard of, as far as I knew. I would address this problem later.

Arlene and I had a lovely flat in Aberdeen, and obviously we would sell it. But we somehow had to get the top book price in two weeks. The more I examined the details of the obstacles, the bigger they appeared, with warts and plooks. But I would not throw in the towel until every legal plot had been put into practice. The first thing I had to do was to check that Arlene wasn't just humouring me until the idea went out of my head, like she does when I hint at buying classic cars and old jukeboxes (she suspects I wouldn't dare go through with it) and that she really did like the prospect of living in a windmill which would no doubt require more than a roll of woodchip and a coat of emulsion to make habitable. Secondly, I had to check that everything we would earn for the foreseeable future could be poured into a bottomless well for repairs and renovations. People would stare, and point. Names would be hurled at us. Like – Windy Miller. And another thing. What did it actually look like now?

Photographs lie. I once travelled to Birmingham with my Dad to buy an MGB GT, as seen in a photograph. When we got to a rundown 'scheme' on the outskirts of Birmingham, the car bore no resemblance to the one in the photograph. It was a wreck, a dog. It had a broken windscreen, two flat tyres and was a different colour. Could the same deceptive imagery take place in photographs of houses? I was only too aware, in a rare moment of common sense, that the last time I had actually seen The Old Windmill was twenty years before in 1973, through the eyes of a nine-year-old boy, and with most of the rest of the world of architecture having been on a slippery slope since then, could this old building have remained intact?

Back at 14 Place Charente (my Mum and Dad's unfortunate address on account of the continued goodwill between Dalkeith and Jarnac, the result being that no-one in the town would ever pronounce it correctly, even with a hint of a French accent; in other words we lived in 'Place Yer Auntie') the family were preparing the traditional and much enjoyed Saturday breakfast of bacon rolls and tea. Bacon rolls at home were something I missed very much, because bakers in Aberdeen did not know how to make proper rolls, or even how to bake, in my expert opinion as a cake consumer. A roll in Aberdeen was a stale hard round flat salty crusty thing, as I discovered during my first week at college there, having asked for 'half a dozen rolls', only to receive a bag of inedible fossils from the local shop. The language barrier and the grumpiness of the shopkeeper at the Mannofield Mace prevented me from complaining and the ducks in the Duthie Park were well fed that day, while I gnawed on a block of mature cheddar. The things that *look* like our rolls in Aberdeen are actually called 'baps' and taste of too much sugar, and are equally indigestible as I learned in my second week at college when I *pointed* emphatically at the required items, still reluctant to wrestle with the Doric language but finally saying, 'six of *them*, please', thinking I couldn't go wrong. Again, horrible taste, fat ducks, thin music student, diminishing student grant. And that is why real bacon rolls at 14 Place Yer Auntie were such a treat.

Having done some research on the subject of The Old Windmill, and with a realistic picture of the pros and cons of pursuing a sale, this bacon breakfast ritual was an opportune time to announce the fact that not only would we be moving house on account of Arlene's new job, but we might well end up closer to home than we had ever thought, if things were to work out. I could tell that Mum and Dad were happy at the prospect, although wise and cautious enough to contain their over-excitement about The Old *Expensive* Windmill. Inevitably, they would be dragged off to Edgehead for a look. And that is what happened next.

In my mind, Edgehead was a long way away. It was out on a limb, not on the main road to anywhere, and the only people with reason to pass through it were those taking what they thought was a short cut to Dewarton and Newlandrig, the only two villages further away from any main road than Edgehead. In fact, Edgehead was a lot closer than I had thought. The information on the schedule was absolutely correct: ten minutes from the new by-pass, twenty minutes from the centre of Edinburgh, five minutes from Dalkeith, yet having the advantage of feeling like a hundred miles out in the country, such was the privacy and tranquillity. So short was the journey, that there was time for only one story, which Dad was prompted to tell in the car when we spotted a sign to D'Arcy on the way.

D'Arcy, to me, sounded like schoolboys-sniggering-up-their-sleeves material, or part of the French connection again, or something out of a Jane Austen novel. In fact D'Arcy was one of the smallest places in the world, and was not really a village or even a house as far as I could see, but it was the site of Esso's first oil-well in Britain, and had been manned single-handedly by my grandfather from the day it opened in 1937 until it closed in May 1966. Dad told us how his father would ride to work on an old BSA motorbike, and when that was off the road pedal power would get him there. The very thought of pedalling up such gradients in the middle of winter from Dalkeith to D'Arcy was unpleasant enough for me even if, according to legend, a billy-can was always brewing for hot tea. After my grandfather died in 1983, Dad showed me some artefacts of his D'Arcy oil-well days including a fabulous engraved gold watch for twenty-five years' service to Esso, some black and white photographs of the site with its oil tanks, sheds, and towering drill, and a copy of the August 1966 issue of the Esso magazine *Newsline* which featured an interview with the old man to mark his retirement after a total of twenty-eight years' service to the company.

In the summer of 1966 the wells were plugged with grout and the area once again became arable farmland. I recently visited the site, one freezing cold but sunny afternoon in October, and I stood alone in the field trying to imagine the sights and sounds of the place when it had been an important industrial landmark. The only evidence that remained was a big steel oil tank with 'D'Arcy' hand-painted on the side. For a good ten minutes I let the wind blow around me while I set my imagination to work in the hope that some ghosts of the past would visit, but it was not to be. Instead, a territorial, rabid guard-come-sheep dog from a nearby cottage came at me, barking its head off, and my hatred of hounds sent me running back to the car, Gnasher in hot pursuit. I took one last look around from the comfort and safety of the driver's seat of the Range Rover. The town was creeping closer as the architectural eyesore of Mayfield spoiled the view from D'Arcy's exposed ridge and the peacefulness of the countryside that my grandfather must have enjoyed.

Less than a mile away, as the crow flies, stood The Old Windmill.

Aitken Hutchison, my grandfather, at work and, opposite, the Esso Newsline magazine of August 1966.

Esso Newsline/August 1966

ESSO'S OLDEST U.K. WELL IS PLUGGED

ESSO'S OLDEST PRODUCING OIL WELL in Britain has been abandoned and will be plugged after almost 30 years of production. The well is situated a couple of miles south of Dalkeith in Midlothian, and is on top of a 750 ft. high exposed ridge of arable farmland. During winter, the site has been known to be completely cut off by snow. The site forms part of the nearby D'Arcy Farm.

In May of this year, preparations were made for the oil wells to be plugged with grout. This cement slurry will harden and seal each well completely. The equipment on the site will be disposed of and the field returned to arable pasture.

Work on the first well began in 1937 when gas was found in some quantity. Drillers came upon oil in sand at 1,734 feet and 1,761 feet down. Five other wells were later drilled, but only number three was a producer, whose initial yield was similar to that of number one. The oil was sold to British Petroleum.

Only two men

Aitken Hutchison, Senior Plant Operator, worked on the site from 1938 to its final day in May. For many years his only companion was Archie Bone, who retired due to ill health in December 1965.

'It used to get lonely up there,' Mr. Hutchison admitted, 'but I liked the peace. It could get pretty isolated in the winter, but I was usually able to get home to Dalkeith on my motor-cycle. I liked to spend two or three evenings a week with the boys in our local pub.'

Rats were a problem and Mr. Hutchison kept a cat on the prowl and this helped reduce their numbers. Mr. Hutchison used to work from eight in the morning to 4.30 in the afternoon and then the site would be left to the wild weather of the hills. A gas engine pumped the oil from the producing well into a settling tank, where the water was drained off and the crude was then put into two 30 ft. by 9 ft. tanks. Gas from another well was piped from the well head to a small shack on the site where it kept a gas fire going for most of the year and was used to heat the kettle for the men's tea. In the shack were two chairs, a cupboard and a desk.

During the Second World War, when fuel supplies were at a premium, several proposals were made to utilise the gas. The first of these was to bottle the gas under compression and then use it for driving road vehicles. The other proposals were to pipe the gas to the local gas works for the enrichment of the town's gas and to pipe the gas to a local wire works for their furnaces. These proposals did not materialise as the geologists could give no guarantee that the gas reserves, available in the field, were sufficient to justify the heavy capital expenditure entailed in these projects.

Another member of the production staff was James Ellis, now an annuitant, who started work at the outset of the drilling operations, and now lives in Edinburgh. He has always kept a 'fatherly' eye on the well. In 1950 he helped in the plugging of one of the wells below the gas strata.

Now there will be only the silence of the lonely farm to hover over the old field, and in the ancient quarries around the site nothing will disturb the peace and loneliness of those first Midlothian oil wells.

The search continues

While Dalkeith oil wells are being plugged, the search continues on land for oil and natural gas in Britain. At the present time, a seismic survey is being carried out in Surrey and Kent to obtain further information about the sub-surface rock formations in the area. The survey will follow approximately the line of the A.25 road between Dunton Green, near Sevenoaks, and Shalford, near Guildford.

The survey will use a vibration technique which ensures a minimum of inconvenience to householders and land. This technique is silent in operation, requires no drilling or drilling equipment, and no exploding of small underground charges. Further details about the survey will appear in a future issue of *Newsline*.

High on the hills of Midlothian stands Esso's oldest producing oil well. This summer the wells are being plugged with grout and the area will again become arable farmland.

10

The last time I had been in Edgehead the Bay City Rollers, Slade and Wizzard were in the charts, Dad had his woolly sideburns, Hibs had a reasonable team and Randall and Hopkirk (Deceased) was my favourite TV programme. Films like *When Eight Bells Toll*, *Those Magnificent Men In Their Flying Machines* and *The Love Bug* were showing at the Playhouse Cinema in Dalkeith, where my Granny worked at the cash desk for forty years, and my Auntie Margaret worked in the office for thirty, and I got in to see all the films for nothing. A new green and white Bedford ice-cream van called 'Shuggy's Buggy' appeared on the streets of Woodburn. Every night children raced each other to get to the front of the queue to buy ten Regal and two 'black men', all to the chiming soundtrack of the first sixteen bars of 'The Blue Danube' as it rang across the town, and still does to this day; same ice-cream man, same ice-cream van.

Even after so long I knew instinctively how to find The Old Windmill. It was a cold but not bitter day. It was dry and calm. We drew up in the car outside the perimeter fence, Mum, Dad, Arlene and me. I switched off the engine and we walked into the drive. The silence was deafening. No traffic. No neighbours. Peace and calm. Only 'weird bird' noises broke the silence every so often. At least I think they were birds. The Old Windmill stood deserted, like a huge machine which had been switched off, seized and rusted. Tentatively, we stood in the shadow of the giant stone monument, walking round it as if it was a huge beautifully wrapped birthday present too big to know where to start tearing off the paper. I felt we were trespassing on private property and that the owner would come and chase us away any minute.

I found it hard to imagine the huge canvas sails turning, for they were long gone, no trace remaining (ironically no wind could even penetrate the trees to get to the sails now anyway). The Windmill stood in almost an acre of mature garden which had once been the pride and joy of the previous owners. More than seventy trees added to the privacy and the fortress fence remained, marking the boundaries, although it, too, was overgrown by bushes and hedges. The four of us walked round the garden in different directions, surveying the house, viewing it from every angle like a landing party from Star Trek. I would not have been surprised to have come across a strange old man, or a talking badger or Hammy Hamster.

Some of our findings were fascinating, and some just what you would expect to find in a house that had lain empty for so long. A broken window, loose copings, rusty rhones, cracks in the harling, rotten wood, peeling varnish. The Old Windmill was in need of painting, pointing and plastering. Drainpipes needed draining; pavements needed paving.

We all loved it.

On the way home I was coming to terms with the fact that no one had poured cold water over my enthusiasm or dampened my spirits, although I was aware that Dad, Mum and Arlene were a little more cautious and a little less impulsive than I was, and I anticipated a debate about the practicalities of this venture in the near future. This was usual though and in a sense I took it that, because no great objection had been lodged at this stage, I had been given a green light to go ahead and take care of business.

4 A Viewing

The festive season seemed to be a quiet season for househunters. I was happy with that. It meant I could get ahead of the game while others skated, sledged, shopped, ate and drank. I was only too aware that there was, to some extent, a race against time, and we had to sell a flat, inherit a fortune and bid for The Windmill before our competitors awoke from their Christmas slumber. I would have to organise a less than inspiring schedule of meetings with men in suits; estate agents, lawyers, bankers, surveyors and executors, and I devised an order of ceremony in the most amateur way, since I didn't know any other way of going about this kind of business.

My plan, on one hand, was well thought out and realistic, but on the other hand it was no more than a dubious and ambitious speculation. First of all, I had to obtain the keys to The Old Windmill, so that we could have a close look inside and out, officially. Secondly, at the same time, our flat in Blenheim Place in Aberdeen had to be valued and put on the market at the highest possible price and advertised as a bargain. Thirdly, a visit to the bank in my best suit with my hair combed for once would be a crucial step because we would have to convince the bank manager that our salaries would qualify us for a 'humungous' mortgage. And finally, skilled solicitors would have to be employed to take care of selling the flat, surveying The Windmill and hopefully making an acceptable offer way below the asking price; this was the part of my plan which I knew was the most sketchy. I knew that there would be no way of affording the asking price, let alone offers over it, but the best we could do was to go through the process step by step, bluffing our way over the hurdles, until we could bluff no more.

We had a merry Christmas and a happy New Year and friends came to Scotland from Manchester to see us. Chris and Lesley stayed for a few days, and unable to contain my excitement, I drove them to Edgehead on New Year's Day, probably against their will, to show them The Old Windmill, from 'a historical Scottish landmark of interest' point of view I think I told myself, as opposed to being audacious enough to say 'this is the house we're hoping to buy.' Chris and Les drove south the night before Ryden and Company opened up after the festive break, and by first light I was virtually camping outside the shop when their chief agent, a Mr Scott, came to open it. This time I wasn't going to be turned away; I was intent on making an appointment with the estate agent, whoever that might turn out to be. With a bit of persistence I persuaded Mr Scott that Arlene and I had to go back to Aberdeen the next day, and we'd like a look round sooner rather than later.

Mr Scott made a phone call, and an appointment to view the property was arranged for two o'clock that afternoon.

Arlene and I turned up promptly outside Ryden's office where our man was waiting for us. I knew I recognised him from many years ago although it took me a while to pinpoint exactly where I had seen him. At first I thought I had seen him invigilating during the 'O' grade exams at Dalkeith High School, but invigilators tend to be much older retired pillars of society from Eskbank, the posh end of Dalkeith, and this man was too young. The Eskbank connection brought me to the conclusion that at one time I had kicked a ball into his garden by mistake and his angry face made an impression on my memory. But that wasn't right either. Eventually it came back to me. When I was very young I used to be dragged off to St John's Church every Sunday, and he was one of the familiar faces of the choir I'd see staring down at me each week as I sat in the front pew driving my toy cars around the contoured terrain of the carved woodwork, reversing in and out of the hymn books. He was quite short and serious, as if he was still hungover from New Year, not quite ready to go back to work. He wore a beard which reminded me of Grubb, the least friendly looking member of the Trumpton Fire Brigade, who was always last down the pole after Cuthbert and Dibble. These childish thoughts had obviously occurred to me in the church in the 1960s, and on seeing Mr Scott for the first time since then, I was experiencing a kind of regression to my formative years. I didn't mention it though.

Mr Scott didn't seem over-enthusiastic about his job, which took me by surprise because I was expecting a kind of used car salesman's approach, only on a much bigger scale, as he was selling a used house. Instead he appeared not to care whether or not The Old Windmill would be sold. He gave the impression he'd been round the block a thousand times before and that he didn't think that we looked a particularly likely couple to be viewing a house like this, and therefore this was a wasted exercise. We followed him all the way to Edgehead where we at least expected his sales talk to begin, but even then he turned out to be little more than a custodian of the keys, as far as we could see. After turning a few mortise locks, the big heavy front door creaked open.

As we walked into the hall the temperature dropped. The smell was typical of a house that had been uninhabited for a long time. The first door on the right led to the big round living room with the Adam fireplace. On the left, a long hall leading to the kitchen and dining room, surrounded by views of the garden. We stood at the foot of the stairs. The banisters had been rescued from the fire-stricken Haddington House, as I was to learn later from Elizabeth Adair's book, and they led up a bright stairway to who knows what. Beyond that we knew that eventually the *pièce de résistance* would be the room at the top, the Secret Room, which had been boarded up for the sixty years previous to 1958. We eventually started with the living-room because the temptation was too much. We unlocked the ancient bevelled door and walked into a bright white round room. The fireplace was the centre of attraction. It was so big that I expected Dick Van Dyke and Julie Andrews to come down it singing 'Chim-chiminee'. The sash windows were quaint, each one giving a view of a different aspect of the garden or woodlands beyond. On the walls hung two pictures of the windmill – one was an aerial photograph taken by some opportunist firm of high flyers in the 1970s, and the other was a watercolour of the house from the middle of the woods painted by Clara Chalmers, the last occupant. There hung in the middle of the ceiling an iron chandelier, and two ornate wall lamps above the fire, each operated by pull switches, perfectly complementing the feel of the place. Although all the furniture had been removed, the original green fitted carpet remained, and some shelves had been made to fit alcoves within the living room's circumference. There was a smell of emptiness and dampness but the building was intact, and already I could visualise next Christmas's tree over by the window.

The kitchen was small but a perfect size for two non-cooks, and to this day I wish I had salvaged the cupboards and kitchen fittings. They were blue and white, made of cheap materials but had that late-fifties-early-sixties charm that's now evident only on period dramas on TV. The curtains had faded in thirty years of sunshine and the kitchen smelled like the inside of an old fridge. The cooker was the Humber Sceptre of cookers, all dials and lights and chrome, and probably not terribly economical to run but, hell, it looked great. The outlook from the kitchen window was as pleasant as it could be when one was standing washing dishes, and let's face it, this was an important consideration, as I knew I would be spending many long hours in that very spot. Tea towels of classic design hung on the period plastic towel hooks, as if time had stood still in the kitchen circa 1960.

The adjoining dining-room was an equally bright, conveniently-sized room with pleasant views all round. At the far end overlooking a rockery there was a ledge just beneath the inner window-sill which was decorated with a variety of coloured ceramic tiles and used for standing house plants on. I was to read how Elizabeth Adair had designed and made this item, although thirty years on I knew I would remove it, if we got the house, to create more space. An enormous glass panelled door opened into a wobbly terracotta patio. There were arches and alcoves everywhere, and in my mind they were places for telephones, computers, recording studios and musical instruments. Behind the front door by which we had entered, there was a small toilet with a primrose cludgy and a sink with silver taps. I pulled the plug on the cistern and the plumbing worked. Now it was time to venture upstairs.

At the top of the staircase there was an impractically small bedroom with a leaded light window and door. But never mind the impracticalities, it was a delightful room and the door opened on to a balcony. There was hardly anywhere to sensibly arrange bedroom furniture, but I felt it was more important to have a good vibe about the room than anything else. We tugged at the Austrian pleated pull cords as different lights flashed on and off, and then we went next door to the bathroom.

This was the most unusual bathroom I had ever seen. Again, surrounded by windows, you could see out to the trees and distant countryside, but no one bar the birds and bees could see in, and there was a stunning vista of the Lammermuir Hills and Soutra to be enjoyed while sitting on the throne. The bath was the star prize in this room, for it was a European hip bath, one which you could sit in like an armchair. I had never seen the like but instantly agreed it was in fact dead hip. The bathroom was spacious and airy and my plans for its final decoration were formed in seconds.

Along the bright gallery-like hall, a big glass door led to the main south-facing balcony with its panoramic views, and then we came to the final door which I subsequently discovered was another piece of Miss Adair's handiwork. The carved wood from a grand piano's music stand was set in the glass panel of the bedroom door. This attractive, though dirt-ingrained, feature opened into the huge circular master bedroom, which was more like a dungeon, apart from its elevation and spectacular views. The sandstone which was so attractive on the exterior of the windmill did not have quite the same appeal on the inside, and it truly resembled a dark dingy jail, but again had endless potential to be homely and unusual. Tucked away neatly at the far side of the room, was a white spiral staircase, leading to the long-awaited Secret Room at the top of the tower. We climbed the sturdy iron steps, finally poking our heads through a crude wooden hatch. We had reached the end of the line, the room at the top. The Secret Room was no secret any more. At first, I was slightly disappointed that no trace of canvas nor mechanism had survived beyond the early 1800s when the sails had been removed, but it was impossible not to be impressed with this wonderful aspect of the house, and I pretended not to notice an ill-fitted wardrobe which was made of plywood, covered in ivy design wallpaper, and which imposed itself halfway across the room that would be one of the main features of the windmill. My axe would be put to work here as a priority, if we got the house. The walls here were stone, painted white, and not dreary in the least, and a rust-coloured carpet covered perfect wooden floorboards. The views from up top were exceptional and the headroom was incredible. Expecting a small attic space, the Secret Room turned out to be huge, and as a special touch, you could walk out on to the flat roof of the extension and see across the whole world, past Soutra and beyond to the very south. The Secret Room would *definitely* be my recording studio. The Secret Room would *definitely* be Arlene's study.

Compact and bijou The Old Windmill was not, but it had so much potential and charm and character that my determination to obtain it was now ferocious. We retraced our footsteps backwards and forwards until it was time for the jailer man to lock up.

Back in Aberdeen we told Arlene's parents (Sylvia and Ian), who already thought I was barking, about our latest venture. Sylvia thought the idea was insane from the start, and I suppose I admit, justifiably so. To begin with, they shuddered and grimaced at the very thought that we would live in an eighteenth century windmill. There was no heating and it only had two bedrooms despite the sprawling size of it. There was too much garden. Arlene and I didn't know a daffodil from a dandelion. The flat roofs would mean nothing but perpetual trouble. They advised us against buying any property with a flat roof, after a bad experience of rain seeping in their own bedroom lay-in (a type of construction with windows built into the roof). It was too far away from Glasgow, where Arlene's new job would be, and it would make more sense to look for a house in a place called Houston or Bishopton or wherever, or so we were advised. We had never had any desire to live in or anywhere near Glasgow but this was not important to the mother-in-law. The work required to be done to the windmill would be extensive and expensive. Right again. And then, there was the actual price. We could never afford it.

Yes, Arlene's parents were correct on every count. But I had heard them describe us as impractical and impulsive before, and as far as I could see this was a golden opportunity for us to live up to such a reputation and throw caution to the wind. The old saying 'To throw one's cap over the windmill' means to act recklessly and impulsively, and in the eyes of Sylvia and Ian the phrase summed us up perfectly. So their cautious approach was given the standard rejection in the nicest possible way and we knew that in spite us ignoring their well-meant and sound advice, they would still be as supportive of us in their good-humoured way, no matter what we did, which is exactly what happened.

Above: The offensive wardrobe in the secret room and below the dungeon-like bedroom.

The original kitchen which sadly had to go.

All this time I had been dying to track down and read *I Bought a Windmill* by Elizabeth Adair who had renovated the derelict mill in a labour of love which began in 1958 and ended in 1962 withThe Old Windmill finally habitable. I had only resisted the temptation to read the book so far because I thought I would be tempting fate by doing so, knowing that if I overstepped the mark of reality, for some reason not unconnected to sod's law, the plan would never happen. But at this stage, with the wheels in motion and at least half a chance of us going to bid for the house, I felt I could allow myself to read all about it. Making my way to the reception desk at Aberdeen Central Library I asked for the book. The library was silent, except for the rustle of two old men reading newspapers and a student photocopying the *Glasgow Herald's* entire appointments section, saving himself the grand total of twelve pence off the price of his own newspaper. About half a dozen other customers were walking around slowly with their heads at right angles to the volumes reading acres of book spines. Then I spotted it. On the top 'Local Interest' shelf at the left hand side about three books in, I found what I was looking for.

I Bought A Windmill by Elizabeth Adair

I dislodged the book from those which had kept it wedged in its place for some years and went through the motions of blowing away the dust and cobwebs as if I had just discovered Dracula's diaries in the vaults of a castle in Transylvania. Flipping open the cover to the first page, I noted the last reader had returned the book in 1985. I could easily have devoured the book there and then, but I thought if I got it home I could relax and take my time and, of course, Arlene could share it. When I got to the checking out desk, however, I discovered that it was not so easy to leave a library. Not with a book, at least. I had no membership. Damn. The main reason was that since I was in Aberdeen only temporarily (it turned out to be twelve years), I always thought there would be no point in going through the bureaucracy of getting ID together. Besides, being a teacher, I had access to my school library. Nevertheless, I was determined to take this book home by hook or by crook, but I was stuck behind the electronic security gate.

Suddenly I was at a Turkish airport with twenty kilos of cocaine stuffed down my underpants, when I noticed behind the counter the familiar face of a chap I think was called John Smith (a name I shouldn't have forgotten) who some years previously used to appear on my radio programme once a month to talk about the new releases in the book world. Our eyes met for a second and John was all of a sudden my oldest and dearest friend. We caught up with several years' news and then I explained my immediate predicament. With no fuss, I was allowed through customs. No handcuffs or guards, just a loan of a book from John at the library. It's not what you know, it's who you know.

'Could I try this with Mr Ogilvy at the bank, perhaps?' I wondered.

The car sped up Rosemount Place, gambled the amber at the Beechgrove Terrace traffic lights and did a Starsky and Hutch turn into Hamilton Place, before slowing to a sensible crawl down Blenheim Place so as not to upset the terrace of older neighbours, all ex-headmistresses who retired in the 1960s, and whose net curtains rustled mysteriously every time I entered or left my house. They all lived with their sisters and whenever I glanced in the direction of their windows I was aware of elderly silhouetted figures stepping back into the darkened Victorian rooms. I ran up the garden path clutching my book, followed by our two hungry skipping ginger toms who narrowly missed becoming the Manx variety by having their tails slammed in the front door in my haste to get reading. It was safe to say that no can of cat food would be opened until the book had been read from cover to cover and I think they knew that. So Smith and Jones (named by their previous owners) each occupied an arm of the couch in the bay window overlooking the street and spent the afternoon intimidating old ladies and poodles who walked past, their beady eyes trained on the moving targets, giving the impression they were ready to pounce and kill.

That day, I remember, was blustery and cold, with sleet showers, and perfect for being inside by the fire. Arlene and I opened the book and savoured every word and every picture, hopeful that one day this fascinating historical house might be ours. As the cats' eyes moved in sync up and down the street, ours did the same across the pages of the book.

I Bought A Windmill was my bible for the next few weeks and the very next day Arlene and I made the journey south to Dalkeith, where we had arranged to collect the keys to view the inside of The Old Windmill for a second time, this time unaccompanied and uninhibited by Mr Scott or any other guide. By previous arrangement Mrs Typist in Ryden's office handed me a piece of old tatty string with two keys on it. One was a Yale key, the other an extremely old mortise job which looked like it would open up Borthwick Castle. We flew to Edgehead, the Citroen performing like the spaceship out of the Jetsons, only faster.

This patch of land was becoming more familiar territory with every visit, but this time it was *official* and we could get inside on our own, *officially*. After a struggle with the locks for a minute or two the big front door swung open.

Mrs Chalmers had bought the house from Elizabeth Adair in 1962, and had lived there until only a few months ago when illness had meant she was resident in hospital. She had two middle-aged sons, one in Hong Kong and the other in London, and they were the executors of the property, responsible for selling it. Despite their own happy memories of the place, they were obviously keen to part with the house, rather than to have it lying unsold and uninhabited for all eternity, many miles away from where they now lived. In addition, the last few months of inattention had taken their toll on both the house and garden. The Old Millstone.

All of this was to our advantage as prospective buyers.

That day, Arlene and I spent an hour or so looking around, exchanging ideas, and imagining how great it would look when it was finished. As we were on our way out, I stupidly decided to check if the burglar alarm was in working order, because I knew that Mr Scott had left it switched off due to perpetual problems with it during the past few months. Having never encountered such a device before, I only *thought* I knew what I was doing when, out of curiosity, I pressed a button labelled '*nic*'. In a split second the peace and tranquillity of rural Edgehead was shattered as a siren suddenly screamed across the countryside and a blue strobe flashed manically. Birds took to the air, animals ran for the hills. We were in a scene from *Bambi*. As we died slowly and painfully of embarrassment, the next scene was from a bad vampire movie – the bit where all the villagers gather in cloth caps and go running up the hill to the castle at the top of the village, carrying torches ready to lynch somebody. I stared helplessly at the bloody burglar alarm, cursing its very existence and prodding randomly at buttons, as if I was ever likely to miraculously stumble upon the right combination to de-activate it. Arlene cleverly pointed out that '*nic*' had its '*pa*' missing and I had pressed the '*panic*' button. By this time I was deaf in one ear and Arlene had wet herself. A bearded villager with a pack of dogs arrived at the gate and enquired about our identity and our intentions. Our lame and ill-thought out reply was along the lines of 'we were probably going to buy the house and the alarm just went off'. Granted, we didn't look like prospective buyers of this fine building, but the chap swallowed the story anyway. This helpful neighbour was unable to switch off the alarm himself, but he knew that someone in the village had a note of the secret combination, so the next hour involved us knocking on all the doors on the village until at last, James Bond punched in the correct code and the wailing ceased with only one second to go before the planet exploded.

The silence was almost as deafening as the initial noise had been, and a substantial crowd had gathered by this time. In fact, it was the biggest crowd I had ever drawn in a musical career spanning ten years. The head neighbour with the beard and dogs watched over us as we locked up the house again and left the premises. Sheepishly, we drove off down the bumpy lane in humiliating reverse gear, all the way to the main road, while the others muttered things about us behind our backs. We were sure of at least one thing. The Edgehead Neighbourhood Watch Scheme worked a treat.

5 Mental Arithmetic

Aberdeen was always a good few degrees colder than Edinburgh. As a student I had always felt a noticeable difference returning to Kepplestone Halls of Residence after a blistering hot summer touring Europe, and a week or so back home in Dalkeith to acclimatise to Scotland while catching up with old school friends was always a good idea. Finally, there came the icy plunge pool of the Granite City at the beginning of October where you could see your breath all day long. Despite the cold, Aberdeen was always bright and sunny on the first week of term, as if the weather had been arranged specially for students returning from their eternal holidays.

Our flat in Blenheim Place in the west end of Aberdeen had been a lucky acquisition. When it was last on the market, a surveyor acting on behalf of a prospective buyer pointed out that the flat was subsiding and should be avoided. This was not particularly surprising news to his client because the adjacent street, Hamilton Place, one of the most desirable residences in the city, was notorious for slip sliding away. In simple terms, the grand and beautiful buildings of Hamilton Place had been built on the Denburn river, and to the horror of so many homeowners the street, in spite of its architectural beauty and hob-nob factor, it now looked like a cartoonist had sketched it freehand with a good drink in him. With Blenheim Place being so close by, the idea of subsidence was not hard to swallow and the prospective buyer backed off, leaving behind the rumour that number 134 was also cracking up. House hunters turned their backs on the flat in their droves, but on the evening we viewed the place, the man who lived upstairs made a point of bumping into us, telling us the tale of woe and waving a certificate from some high flying buildings surveyor with *more* conclusive proof that in fact the house was *not* sinking. We were delighted about this of course, but realised immediately the friendly neighbour's ulterior motive. If the flat underneath his own was condemned, what chance would he ever have of selling? Having heard two sides of a story and wanting the neighbour's one to be true, Arlene and I had the house *doubly* surveyed, insisting on cast iron guarantees, drawing attention to the rumours and the certificate. In the meantime, all other interested parties had been put off completely, and the one party who had made an offer 'subject to survey' had pulled out. By now, the couple who were selling Number 134 were panicking about the subsidence, and it seemed they were quite prepared to take less money than they had at first hoped to get, and we were prepared to make an offer that very day. Arlene and I were in exactly the right place at the right time with the right money, and before any of the other prospective buyers could give it a second thought, 134 Blenheim Place was ours. Jammy.

The flat needed a great deal of work but it was mainly cosmetic and decorative. On the day we received the keys, the owner wished us well and left with a bag of his late mother-in-law's possessions which had remained in her flat all this time. At lunchtime on the same day he realised that he had accidentally removed the telephone, which belonged to BT, and he returned to the flat to give it back. To his obvious horror, in the short period of only a few hours, an enormous skip had already been filled to the brim with most of the inside of his late mother-in-law's house, including the dismantled kitchen, two 'modern' fireplaces, the entire bathroom suite and half a ton of hideous curtains and carpets.

In the three years that we lived in the flat in Blenheim Place, we transformed it into a very desirable dwelling house at not much cost, and with house prices going up and Blenheim Place remaining a desirable area, we were in a fortunate selling position at a time when we needed a quick sale. However, I must admit that the issue of debatable subsidence did come back to haunt me. If such an ambiguous situation was to arise again, it would be goodbye Old Windmill, hello Blenheim Place, for the rest of our days. We arranged to have a valuation undertaken, and hoped that this would also reveal to us any problems which a surveyor might pick up; or *not*, preferably. The Blenheim Place evaluation was going to be the first item to cost us any money, and from this moment on every penny mattered. So off I went seeking 'favours'.

A very good friend called Brian Hendry had for some years been a successful businessman in Aberdeen and had contacts you wouldn't believe. Brian knew many famous and influential people from film directors to world class footballers. He rubbed shoulders with Jacques Shirac, Rod Stewart and Craig Brown, and always drove a flash forty grand car. Yet there was nothing dodgy about Brian. He wasn't a mason or a gangster; he was unassuming and just genuinely a good Samaritan. In the summer of the previous year, Brian had asked me to provide music for a cabaret which was being run in a function suite in Aberdeen and for some reason, whether for charity or for fun, some of his closest friends were going to be performing to a private audience of people they knew. I was asked to accompany and rehearse the acts, and play on the night of the performance, attempting to hold it all together musically. The acts were unusual to say the least. Two of Aberdeen FC's and Scotland's star players were Alex McLeish and Willie Miller. They would be performing an Eric Clapton song with Alex dressed as an amplifier in a huge cardboard box with a speaker painted on the front and head and limbs sticking out the sides, and Willie as a guitarist, with an inflatable electric guitar plugged into the 'amp'. Willie sang the words and Alex made twanging noises to represent the famous guitar theme. I was their 'musical director' but the incident wasn't one I would put on my curriculum vitae. Another act on the star-studded bill was 'The Three Tops'. Brian was one of the Tops, as was his brother Bev, now a big shot in America (though not as a singer) and the third Top was a gentleman called Jimmer, Aberdeen's top solicitor/buildings surveyor. You can see where this is leading. The Three Tops pulled 30 denier American Tan stockings over their heads and sang 'My Girl' and 'Hit the Road Jack', convincingly, choreography *et al*. A splendid time was had by everyone and to cap it all the buffet was delicious.

The point of this story is that in exchange for my musical services that night, Brian was able to return the favour by putting us in touch with Jimmer, who was the man we needed to value the flat in Blenheim Place and, if possible, jack up the valuation to the boundaries. In other words, to err on the high side, if there was room for such ambiguity in the world of valuations. Brian was sure Jimmer would be able to help.

I think Jimmer was disappointed that my phone call was not in regard to booking 'The Three Tops' for the Aberdeen Alternative Festival, but in fact related to the more mundane everyday task of house surveys. Jimmer said he would oblige, which he did, and his valuation of the flat was almost exactly what we had hoped for, without Jimmer having to 'err on the high side' or soil his reputation or break any code of conduct in any way (which frankly I would have been happy for him to do at the time). Jimmer really came through for us and to top it all he wouldn't take any money. Whether this was because it was time for a charitable gesture, or because he wanted me to keep *schtum* about his extra-curricular activities as a soul singer with a stocking over his head, wasn't clear. The good news was that, after Jimmer's valuation, 134 Blenheim Place could now be advertised for sale in the property register for exactly *double* what we had paid for it three years earlier.

I reviewed the situation so far. Phase One of my plan had involved viewing the windmill and checking its viability as a dwelling house, Phase Two had just been completed, and Phase Three was to actually sell 134 Blenheim Place – in one week.

The phone was red hot. Tourists came and went. Some said they'd get back to us but never did. Others said the bedroom was too small. Some wanted desperately to buy the flat but couldn't sell their present place; they were stuck in a way we didn't want to be. I had never seen such a tidy flat. Not only that, it remained tidy for several days in succession. I felt like a tour guide for the National Trust. If we had charged eight pounds a head from everyone who came to view the flat in Aberdeen, we could have paid cash for The Old Windmill. Many faces and characters passed through the hall, into the tiny bedroom, admired the wee bathroom with the beautiful stained glass window and the corner bath, the queer-shaped dining room with the recording studio, the long galley kitchen and the big living room with the bay windows. From music therapists to oil industry roustabouts, nurses to nightclub owners, they came and went.

Then at the end of the first week of 134 Blenheim Place being on the market, an offer came in for *exactly* the amount at which Jimmer had valued the house – our exact asking price – the price that would complete Phase Three of the master plan.

Subject to survey.

Damn. At times like this the pessimist in me comes out. I was now sure that the foundations were floating down the Denburn. Dampness was climbing up the walls and the roof had to be replaced. I sat and sweated it out until we heard from the men in suits, who always move in slow motion.

We left Phase Three hanging in the balance for now and moved on to Phase Four – a visit to the bank – in which Kenneth dresses up in a suit.

The Tullos branch of the Bank of Scotland was situated in the middle of an industrial estate on the outskirts of Aberdeen. We had a particularly good relationship with this bank because a neighbour of Arlene's parents, John Reid, used to be the assistant manager there before taking early retirement. Not only did he agree to fund my first mortgage for my very first flat, allowing me to stop living in squalor like Romany Jones in a two berth caravan in a friend's garden on the Keith Hall Estate, but when the bank was being refurbished he offered me thirty square feet of floor tiles to furnish my flat. The unusual deal was that I had to go and collect them myself at some unearthly hour, when the bank's renovation work was being done, and John would give me a hand. So one night, around midnight, I drew up in my Marina Estate, which looked like a stock car after an accident in Perth some months earlier. I was wearing old clothes in anticipation of a messy job and black gloves for fear of breaking a nail. John helped me load acres of floor tiles from the bank into the back of this dodgy-looking vehicle. All the time I wished a police car would arrive, or an old passing busybody would report us, because although the scenario had the appearance of an armed robbery, we had a lovely excuse and I was dying to use it. Had we actually been robbing the bank without an excuse, I am sure an army of police officers on night manoeuvres and the Neighbourhood Watch Committee would have arrived at the scene simultaneously, such is life.

John took *very* early retirement *very* soon after our midnight mission and he went off to Inverness to become a photographer, although it has since occurred to me that perhaps floor tiles were not the only items removed from the bank that night, having learned that John spends most of his time golfing nowadays. Anyway, the new bank manager was Mr Ogilvy. He sat proudly in his refurbished office in his refurbished bank with its new carpet and he seemed far too friendly to be a bank manager.

Being in Mr Ogilvy's office was like being in the headmaster's office at school. He looked relaxed in a big plush chair across a huge desk which, for a bank manager, seemed awfully tidy and empty, bar a family photograph, a neat leather bound folder (which Leonard had brought in, virtually on a silver tray, before retreating with a polite bow) and a brass name plate which read 'Mr. J. Ogilvy'. The new carpet which had replaced my brown carpet tiles was a grey and red mottled design which I presume had been commissioned in an attempt to jazz up the atmosphere, and it had been colour co-ordinated to match the blue/grey wallpaper, the *Space 1999* door handles and the office furniture. We sat at the foot of the big desk on small seats which I am sure were chosen to make people who were begging for a bank loan look as though they were doing just that, in stark contrast to Mr Ogilvy's throne, a chair which Jimmy Savile OBE would have envied. We sat with our knees up to our chins, feeling subservient, incapable of assuming a dignified position, no matter how hard we tried. To add to my personal discomfort I was wearing my least favourite attire – a suit, and although to the average onlooker this would not be a strange sight at all, when *I* looked in the mirror or caught my own reflection in the bank's tinted windows I could see Coco the Clown: red nose, big shoes, kipper tie, baggy suit with oversized checks, the whole deal, and circus music and old-fashioned car hooters rang in my ears.

Talking of oversized cheques, our purpose in life that day was to ask the Bank of Scotland for the biggest mortgage in the history of the world, at least that's how I saw it. I had already calculated that we could qualify for the required sum, legally, but we had to strike while the iron was hot. Let me explain.

The bank would allow any couple to have a mortgage to the value of up to three times their joint salary. The whole point of our house move was that Arlene had been offered a job with Scottish Television, and we knew her new salary would qualify us for a big mortgage along with my teaching salary. However, Arlene's TV contract, we both suspected, would not be worth the paper it was written on and was more likely to disappear in six months than in the ten years talked about by the TV executives. I would then have to supplement my school teaching by playing the piano in some of Edinburgh's best paying piano bars who had miraculously offered me cast iron contracts well into the future. I would be playing the 'Girl from Ipanema' until I was seventy-five to pay for this house.

It was a day of big sums, of facts and figures. There was a great deal to discuss but our track record had been good (apart from our usual disorganised financial situation on a day-to-day basis) and we had never missed a mortgage payment. Other things in our favour: I had banked with this branch for some years now, good old John Reid had been a friend of the family and Arlene had a nice smile that could charm the birds out of the trees. In addition to that I believe Mr Ogilvy was genuinely interested, subject to survey, in the prospect of his bank (let's face it) owning an old windmill. As we were such a charming couple with 'a potentially high earning capacity' (yet to be achieved), Mr. Ogilvy agreed to grant us the mortgage, subject to survey again, and Phase Four of my most unlikely scheme ever had been completed. Little did he know that we might well have to sell home baking by the roadside or go round the social clubs as a magic act – The Great Skinto and the Lovely Arlene. Otherwise the plan was going well, and now the final stage was to bid for the Old Windmill; and then move in.

Two weeks after selling my soul to the Bank of Scotland, the piano bar jobs dried up, thank God, and the 'Girl from Ipanema' caught the 'Chatanooga Choo Choo' (for all I cared) although fortunately other opportunities in the world of radio broadcasting presented themselves to supplement the lost income. Additionally, nine months later, (not far off what we had predicted), the department in Scottish Television that Arlene was working for fell in to a state of mismanagement and disorganisation and her contract ended. *Quel surprise*. For nine months she found herself 'on the brew', until a lovely opportunity with the BBC came along just in the nick of time.

The point is, that on one particular day in history, the day of our visit to the bank, all our affairs had to be in order – salaries, contracts, security, track records, prospects. And because of the favourable circumstances on that very day, we were able to stick to my big stupid plan, hanging on by the skin of our teeth, with the ownership of The Old Windmill becoming more of a reality than just a pipe dream.

You drive into the village of Edgehead on a hot summer's day, coming from the Whitehill end just off the main road. The first thing you notice is the breathtaking view from the brow of the hill; the unspoilt countryside, fields of yellow, gold and green as far as Soutra's meeting with the horizon. It's like a view from a roller-coaster. You can see right down the straight main road of the village, Dere Street itself, to the hamlet of Ford, then up again to the red roofs of Pathhead (which look their best from such a distance), and then up and away over the hills to the sky. On your right, children play in an unspoilt park; swings and roundabouts, chutes and see-saws, all in bright clean primary colours. Three old men with no agenda, dressed in brown Fred Perry shirts and tanned as if just back from a fortnight in Majorca, reminisce on a bench while their tethered dogs rest in the shade underneath, out of the tiring sun. There's a family enjoying a picnic over there, miles away from their tenement in the city, and some wee boys are kicking a ball around the big pitch where my football career began and ended. You head for the shade that's thrown over the edge of the park by the still trees and you sit on the stone wall that's been there since before your mother was born, next to the hot timber bus shelter that smells of creosote in the baking sun. A car passes by every ten minutes and a bus every six hours, but you hear them coming from miles away, a long crescendo from the other end of Dere Street, the Romans' unswerving master work that led you here. The flowers at your feet are in full bloom, and the village is draped in them from top to bottom; rose beds, hanging baskets, herbaceous borders and decorative barrels and tubs. Every colour you can think of is there. You note the silence of a Sunday morning, though it's a Saturday afternoon, and you stroll to the bottom of the village accompanied by the birdsong soundtrack: various artistes. Heading through the forty shades of green in the copse on the left hand side where it's much cooler, you walk as far as the cemetery, set in its own peaceful silence at the foot of the village, and then back up the hill again past the coloured cottages to the park, where the young family are being rounded up and shepherded into the people carrier before heading off back to the town. Looking through the bars of the kiddies' climbing frame, you notice away over there in the top corner of the park The Old Windmill watching you through the trees.

... No traffic. No neighbours. Peace and calm...

One Cap over the Windmill

You would never know it now, but Edgehead was once primarily a coal mining community with the first official mention of the industry appearing as far back as 1743, when the 'coaliers' (as they were usually referred to in the official parish records) lived in the farmhouses. A quick history lesson. Blackside Farm was the first area to be mined in these parts, setting a trend for a move away from exclusive farming to prospects in coal, for some at least. There are also records of the residents of Cranstoun, Sauchenside, Chesterhill, and Edgehead, relating to population, occupation and occasionally more detailed information. From these records, it is easy to see the growth of coal mining from its inception in the 1740s to its further development well into the twentieth century. The documents also tell us that sixteen miners were resident in Chesterhill in 1791, with names like John Penman, John Thomson and William Russel, and this number grew steadily over the next hundred years. By 1840 there were twenty-two mine shafts operating within half a mile of Chesterhill, including those named Brummel Pit, Civility Pit, Gig Pit, Belham Pit and our own particular favourite for obvious reasons, The Windmill Pit.

Records also show that those who worked the black seam eventually tended to live at the bottom of the village of Chesterhill on the right hand side, while at the top on the left (primarily in the 'square') there lived the tradesmen: the stonemasons, the carpenters, the millers. In the original statistical account for Scotland in 1792, two millers are registered as being resident in the parish of Cranstoun (sometimes spelt with a 'u', sometimes without), and it is thought that Chesterhill Mill was their place of work.

Throughout this industrial revolution, farming still thrived on countless sites in the area: Sauchenside Farm, Blackside Farm, Coldwells Farm and Edgehead Farm among others, many of which are still in business today.

I had known for years that Edgehead, as I called it, was also known as Chesterhill. The school in the village was called Cranston Primary School and I just presumed the village had once been called Cranston. Then the farm opposite the school was called Sauchenside Farm and I had seen this name on many of the old maps I had found, so it must have been a great empire at one time; and just in case there weren't enough options already, the old maps often labelled the village 'Chesterhall' or even 'Chester Hall'. Sometimes Edgehead became 'Edge Head' on the maps too. There was obviously some confusion here, if only in my tiny mind.

I tapped the knowledge of the village historian, Bill Davidson, who seemed to be passionate about the history of the area, and regularly organised history evenings in the village hall. We spent an afternoon up to our knees in his notes with maps strewn all over the place until the mystery had been solved and it was all clear in my head:

At the bottom of the hill, the parish of Cranstoun (according to the Register of Births) existed as a thriving community with a small but steady annual birth rate, with the church and the school at the hub of the village. Then suddenly the Register of Births ceases to credit Cranstoun at all, while at the same time an increase is noted in the population of the nearby villages of Sauchenside, Chesterhill and Edgehead (each consisting of barely a few houses). The expert opinion is that everyone moved away from Cranstoun altogether (perhaps for reasons of change in job opportunities, who knows?) albeit just a few hundred yards to the closest neighbouring hamlets, but this does explain the demise of any reference to Cranstoun ever again. The name, however, is kept alive in the village today as the school's appellation.

Chesterhill we know to have been the name of the cluster of houses forming the square, followed by the other cottages in the village on the left looking down the hill; although, only a couple of hundred yards away at the other side of the road, the windmill was *always* known as Chesterhill Mill. It's only the mis-spellings on various maps that make Chesterhill's pedigree a bit dubious.

The village of Sauchenside appeared to be named after the farm at the bottom of the hill on the right, or perhaps it was the other way round; but as an independent entity Sauchenside seems to have been phased out as the name of anything except the farm and one of the cottages. There's no real way of telling what actually happened here, but *most* people who live there nowadays say they live in Edgehead, perhaps for the sake of convenience.

Then there's Edgehead itself. This is the name of the farm on the brow of the hill, and it's the first name to be seen on any sign before you get to the village. Originally the farm was called Bennet's Farm, first operational in 1758 or before, but it changed to Edgehead Farm in 1785, and this is the name that seemed to stick with the locals. So even Chesterhill, Sauchenside and Cranstoun, in time, came to be known as Edgehead, and Edgehead is the name that prevails to this day, appearing on the road sign in the biggest and boldest letters, with Chesterhill second on the bill in brackets, just like having the Beatles followed by The Kinks on a poster outside a concert hall. And what of Cranstoun and Sauchenside? The billing equivalent of Gerry and the Pacemakers?

I sat in the car outside the Odeon cinema in some back street in Aberdeen, having left early after bringing to the attention of the manager that beyond the Art Deco foyer with the marble steps and lovely carved doors, the cinema itself was a freezing cold, tiny, uncomfortable dump, not conducive to a night of cinematic entertainment. There had been problems with the sound being nearly two seconds out of synchronisation with the pictures and I was irritated by this to the point of politely asking for a refund halfway through the film. I was eating a haggis supper from a dreadful chip shop opposite, where *deep-fried* pizzas were normal fare. The rain belted against the window as I devoured a less than nutritious meal, cursing every chip for being saturated in grease, and scowling at Northsound's output on the car radio; this was the radio station I used to work for until a change in management left us in the hands of, in my experienced view, an incompetent baboon, name withheld. Every so often, in a perverse kind of way, I still tuned in to Northsound just to confirm how much I couldn't bear to listen to them anymore. I liked Aberdeen very much but I was now giving myself an ever-increasing number of reasons for going back to Edinburgh, or thereabouts. The things that had attracted me to Aberdeen in the first place no longer mattered quite as much, and I missed my old stomping ground more and more. Reconnaissance missions for a place to live had by this time taken us to the States and to other parts of Europe, and although travelling was still a major attraction for me, the 'home is where the heart is' cliché rang true as I approached my third decade on Planet Earth.

Arlene and I were very close to her family, but she too recognised that career-wise Aberdeen was a very small pond in which she had quite easily become a big fish. The challenge of a wider market in broadcasting was an exciting prospect, particularly as she had grown to love our regular trips to Edinburgh, my family, a more cosmopolitan atmosphere, and salt and sauce on her chips. She was ready to leave behind a job in a TV station where the management were beginning to take her talents for granted. Of course there were some things we would both miss about Aberdeen, as we do now; but as I sat there that night, up to my knuckles in lard, there was a real buzz about the thought of a new start in a new place, or even an old place. In truth, even if The Old Windmill *was* to pass us by, the big picture still looked exciting. A flock of seagulls got to finish off my supper, more fool them, and I plotted my way back to Blenheim Place in the black rain, windows steamed up and a layer of fat obscuring the windscreen even more as I tried to clear a hole to peer through.

When I got in, Arlene was sitting at the Grampian newsdesk in a nice outfit reading the late news on the telly as impeccably as ever. I sniggered to myself as I knew that the unsuspecting viewer could not see that under the desk Arlene wore tatty jeans and slippers which were big rats' faces with sunglasses on, and to shatter the illusion even more, her ironing board stood in the corner of the studio (more of a cupboard than a studio) as she ironed her clothes in between the programmes. What a glamorous life. I flicked around using the remote, only to find nothing even vaguely interesting on any of the channels so I switched off the TV and gathered my thoughts about the next phase of the plan, which until now had gone well. I looked at the list on my crumpled scrap of paper.

Next, we would meet the man from Ledingham Chalmers, Solicitors, to talk about bidding for The Old Windmill.

Arlene, who doesn't seem to worry so much about issues of property conveyance, was absolutely right not to worry. The sale of 134 Blenheim Place was moving along quite nicely and Phase Three was nearing completion, but the solicitors' part seemed to take forever. All they had to do as far as my birdbrain understood it was to sell the flat for us to the person who had already agreed to buy it. How hard could that be? I remember thinking that if their lives depended on it, surely they could do the necessary paperwork for your average house sale at twice the speed. I was convinced they were inventing documents and procedures to spin the job out a bit. This unsubstantiated view was only my personal, uneducated, irrational, gut feeling and I don't think I would have ever been so bold as to suggest this hunch to the legal fraternity. However, our bill for their part in selling our flat seemed an extraordinary amount and I would not have been surprised to find the price included a free car or a holiday in Majorca. The money from the sale – if it ever arrived – would go straight through our fingers into the bank along with some paltry life savings which might have one day stretched to a weekend in Stonehaven, and some funds from a list of items which we would have had trouble offloading at a car boot sale. They were:

A Tascam Porta One (one of the earliest four track recording machines, now quite desirable and this one in particularly good condition at the very attractive price of £60).

A Roland Juno 60. (This coveted old diehard of the gig comes with flightcase weighing twice that of the instrument and all faders work perfectly. Steal it for £100).

A Tanglewood acoustic twelve string guitar. This was actually the less popular nine string model owing to one of my pet hates being to change or renew guitar strings. (Finished in sunburst with an ornate scratch plate, this beautiful item comes complete with plectrum, strap and tartan carrying case. Lovely tone at only £45.)

Spacehopper. (Relive the magic of the seventies with this nostalgic bouncing toy. Finished in classic orange, every household should have one. Hop down the street for only £20.)

Three Triumph Dolomite Sprint Wheels. Don't ask. (Be the envy of all the others in the Reliant Robin Owners' Club…)

and so the sale of the century went on. To my amazement people actually came with their money and took these ridiculous items off my hands! The only things which had to be prised away from me and caused tears and tantrums were my vehicles. The old Honda 550 Four K which I had bought from a sixty-five year old minister only a year earlier had to go. It had been a cruel winter and the bike had been garaged for most of the time. I was no longer able to justify hanging on to it and a dealer in East Lothian eventually took her off my hands. And the big red sleek comfy spacious aerodynamic fuel-guzzling beautiful quirky troublesome Citroen CX was replaced by an ancient pug-faced Peugeot 205, which farted along the street at forty, tops, but delivered a pocket-pleasing thirty-five to the gallon. I was up to my armpits in the economics of affording a windmill. I was turning into my Granny, bless her heart, who had worked out that if you took a bus to Fine Fare in Loanhead (free for old age pensioners, as they were called in those days) you could save five pence on a tin of beans, the same beans being five pence dearer in St Cuthberts in her home town of Bonnyrigg. And it was a day out.

The kitty was as full as it was ever likely to be, bar selling the matching cats as an exquisite pair of ginger slippers, and the clothes we stood in. We were now poised ready for the final step: to make an offer that we could afford. We had pushed the boat out to the mid-Atlantic and the day of reckoning had come. Mr O at the bank was happy for us to borrow large amounts of cash, Mr Brodie our solicitor was happy to take a sizeable chunk of it, and we were about to offer all the money we had and all the money we could borrow in exchange for The Old Windmill, Edgehead. With more neck than a giraffe, and I think to the embarrassment of Mr Brodie the solicitor, we offered a sum of money which fell some thirty-two thousand pounds *short* of the minimum asking price. Then we waited for a response.

It came as no surprise to us when the reply was 'no'. For this simple and succinct answer, I recall, we had to wait several days, due to the speed at which solicitors and their correspondence travel. Conveyancing was costly. If solicitors, and all the things their business entails, were modes of transport, they would be rickshaws. If they were fuel economy, then they would be Range Rovers with holes in their petrol tanks; and for this reason I knew that there would very shortly be a limit to how many times we could afford to bid for the windmill, stabbing blindly in the dark, gradually raising the offer until the executors accepted one of them, hoping that they would be desperate or at the very least glad to sell the place.

I thought it would make more sense if we could somehow find out through the grapevine the *exact* figure that the executors would accept, then of course we could either offer that very sum, or abandon the idea completely and go and live in a sensible house like everybody else. Solicitors have their codes of conduct and although I am sure that nothing sinister goes on, I have my suspicions about how much they know in relation to how much they are actually prepared to tell a client. So with this new completely unfounded paranoia in mind, I temporarily followed another trail while Mr Brodie wrote letters in slow motion.

Jimmer had worked wonders on getting the flat in Aberdeen sold. He was a top professional. In fact he was a Top. Jimmer's parallel career as a karaoke king was clearly a source of embarrassment to the man, and when I phoned him up I got the impression that he was squirming at the thought of me publicising details of his alter ego on my radio programme. Nothing could be further from the truth. I was quite comfortable with Jimmer's alternative career, and I would encourage this sort of thing. I knew one headmaster who was moonlighting as a sound engineer, and another who had a double life as a football manager. I knew an Inland Revenue inspector who played bass in a soul band at weekends and didn't declare the wedge, and a grocer who by night was a male stripper; and I was an acquaintance of a top boarding-school master who enjoyed seasonal work on the Waltzer at a well-known residential funfair at a popular beach resort down the coast. Scream if you want to go faster!

So Jimmer was absolutely fine in my book and I did not even hint at the 'do me a favour or else' approach, I swear. I simply explained the situation and our new predicament over hitting the right price target. Within hours, Jimmer was on the phone with a cast-iron, money-back, no-quibble guaranteed figure that would be acceptable to the executors of The Old Windmill, as divulged by God knows who from God knows where, but frankly that didn't interest me much. The figure was a good deal more princely than the first offer we had made, but still an awful lot *less* princely than the original asking price, much to my surprise. The sellers were prepared to accept fifteen thousand pounds *less* than the now legendary original asking price. We now had concrete figures to work with, and if you're still with it and you got your Arithmetic 'O' level, you will have worked out that we had to find, steal, borrow, launder, whatever, another seventeen thousand pounds from out of thin air. The prospect of busking on Princes Street loomed. Only I'd sold my bloody nine-string guitar to a hippy in a psychedelic Volkswagen caravanette, and he was probably raking it in outside Boots the Chemist.

I remember the situation becoming more and more like a game of Monopoly. It no longer mattered how much money we actually had either in our pockets or in the bank, it wasn't ours anyway. We were now on the Planet Debt, and at this point we knew that if this newly discovered unequivocal sum of seventeen thousand pounds could be raised, then the windmill would be ours, subject to survey and legal surprises. We felt we had come such a long way, that to drop at the last hurdle would be something to be regretted. I tried to convince myself of this by looking at how I had struggled to pay for my first bachelor pad (after residing in an old caravan for a year with no running water, no heat and no light), eventually selling it for twelve thousand pounds. Then our first flat together (this one actually had a bedroom in it) which cost the earth by comparison and involved the same amount of anxiety, soul searching and pound stretching. Only three years down the line I recalled that the monthly mortgage payments on the flat in Blenheim Place had become relatively negligible. I also tried to justify the sense of taking on a big mortgage to myself by comparing our prospective monthly outlay to the current cost of renting flats in the centre of Edinburgh, cottages in the country and semis in suburbia, and to my delight our option came out most favourably. It would be at least as cheap to have this wonderful patch of our own, as it would to hand over all our money to some greedy bent landlord, getting little in return, apart from a temporary roof over our heads. Other points for discussion were :

One. Another windmill so close to Edinburgh, set in an acre of mature woodland would *never* come on the market again in our lifetime.

Two. If we procrastinated any longer Lord and Lady Muck could well be driving through Edgehead from their residential estate in Sussex while driving young master Henry to Rannoch School for the start of the new term, spy the windmill for sale and pay cash for a super little holiday home in the shape of an eighteenth century windmill. What a wizard wheeze that would be down at the bridge club.

Three. It was the height of cool.

I found myself dressed in a silly suit again, feeling like everyone was looking at me as we made our way to Mr Ogilvy's nice bank in Aberdeen. The process was more swift and straightforward than we ever could have imagined. Mr Ogilvy had become 'Jim'. He was eternally interested in the sale of the old flat, our progress in the pursuit of The Old Windmill and our general well-being. He was an easy man to ask an extra seventeen thousand pounds of. His manner on that day was pleasant and professional to the last. Enquiries about our health and tales from his own childhood peppered the conversation while he searched the previous application forms, presumably to check whether or not he had it in his heart, or vault, to allow the extra money. Without even letting us sweat or squirm he said it would be fine, and Leonard returned smiling within five minutes with new forms, on which we signed our lives away. It was that easy. We would never see, smell or touch the money, as it travelled at the speed of light into an account which was already so overdrawn on paper that the minus sign was about three columns out of alignment on the statement card from the autoteller!

6 Sold

It was the middle of January 1993, D-day (Debt-day) was approaching, and at last I felt we had surmounted most of the obvious problems. Although at school I had failed miserably to add mathematics to my list of qualifications after three half-hearted attempts in the face of monumental boredom, my teachers might have been proud if not amazed that such a big hard sum went through this particular head and came up with the right answer in the end. We were now undoubtedly in pole position by a mile (barring any last minute Ruperts with pots of cash), still gathering pace, momentum, excitement and adrenaline, and the moment had come when we could skip into Mr Brodie's office at Ledingham Chalmers and instruct him (legal jargon) to offer on our behalf 'a considerable amount of money' to the executors, two brothers by the name of Chalmers (another coincidence) for the purchase of The Old Windmill, Edgehead, knowing fine well that the figure would be acceptable to the Chalmers brothers, thanks to Jimmer's Huggy Bear style information.

Why this could not have been done over the telephone I will never know. But in the hands of solicitors we were walking in treacle again. We had gone from 78 rpm to 33 rpm. The process had gone from being Donald Campbell's *Bluebird* to a nun on a bicycle, and all we could do was wait. We loitered around, waiting for two days for a simple answer.

We had hoped for an outright 'yes'; as opposed to a 'yes, but'. And you will hardly believe the complication which now presented itself. By some ancient law, if Chesterhill Mill was ever to be sold, then it should be offered first to the present Earl of Stair, on whose land the windmill was originally built. He would then be given the opportunity to repossess the building 'for the sum no less than the highest offer made to the seller'. Well, for all I knew the Earl of Stair could be living in Egypt with no desire to set foot in Scotland again. On the other hand my pessimistic side told me he was rubbing his hands with glee for at last, after thirty years, he could reclaim the mill, which had been associated with his forefathers, to turn it into a country craft shop and tearoom called Flora's Kuntry Kabin. But before we would know for sure, the Earl of Stair had to be tracked down, bombarded with unsolicited legal documents, prodded to respond to them and persuaded to sign away the building until next time. Hopefully. We waited impatiently.

I cannot recall the exact moment when it was time to jump up and down, or crack open a bottle of champagne or shake hands or hug everyone, like at New Year time or at the finishing post in a race. I cannot remember actually waiting by the telephone for hours, or hovering seedily outside the offices of Ledingham Chalmers in anticipation of the result. But I do remember that the good news came from a colleague of Mr Brodie (who had himself gone off on a skiing holiday mid-deal) that our offer had been accepted. We did it! We bought a windmill!

The Old Windmill was officially ours on January 27 1993, and I, for one, could hardly believe my luck. On that cold afternoon, Arlene and I stood among autumn's long abandoned leaves in the shadow of the huge tower. Looking fantastic against a blue-sky backdrop, it still dwarfed me as it had done when I played football for the cubs twenty years before.

An entry date of March 12 had been agreed, when the keys would be ceremoniously handed over. The timescale was carefully chosen to allow ourselves a realistic period in which to sell the flat as we awaited the completion of the legal paperwork and all that entails.

It was suddenly all a bit unreal. While there was scope for so much celebration and happiness I kept thinking that it was too good to be true, and that someone would trip me up as I approached the gate, or the building itself would sink and crumble into one of the disused mineshafts which lay somewhere below. And so with my most unlikely scheme having been accomplished, any serious cork popping would be postponed for the time being.

Dad was delighted, and all of a sudden he became a key player. The very next day, January 28, he made contact with various tradesmen to start work at the windmill. March would be upon us soon, and we wanted everything to run smoothly, so forward planning was essential. We invited quotes for the installation of a central heating system as a priority, as this was bound to cause the most mess and disruption. Hamilton Brothers of Dalkeith, Colin Dickson and Billy Hosie were all in the running for this job, among others.

I unearthed a small tacky five year diary quite recently. My sister had brought it back as a present for my Mum from Guide Camp in Linlithgow in the summer of 1979. The only entry in the diary was made by Mum on July 28 of that year. It reads:

> *Received this diary from Jacqueline from Linlithgow.*
> *Mother and Jean go to Blackpool.*

Dad had obviously found it in a drawer fourteen years later, wrote on it the working title 'We Bought a Windmill', and put the book to better use, taking notes of our developments as he went along. Fortunately, Dad's diary allows us to keep track of precise dates and costs, as well as preserving some of the early memories of the first few months of 1993, just as Elizabeth Adair's diary had done more than thirty years before. Dad's diary also contained lists of names and telephone numbers and their respective trades. He spent the next few evenings on the phone trying to pull together the team who would renovate the house. Some potential workers were already booked up well into the future and some were on holiday. Others priced themselves out of the game. But the search went on, and within a few days Dad had booked Lionel Thomas for an asphalt repair to the balcony, R. Main was employed to fix the slates on the roof of the tower, Willie Park the plasterer was standing by for action, Rab Purves the brickie would be one of the first on the scene, Ron

Rankin from a cottage near the village came to upgrade his original security system and John Noble, the spark from Bonnyrigg, would do some rewiring. Others came along to poke around and give us their quotes, but for various reasons, mainly financial, they were rejected.

On the 29th of January, only one day later, the good news came that the sale of 134 Blenheim Place, Aberdeen, was finally complete! I was only too aware of the big daft sum that had been going on in my head all along, and how things could still go pear-shaped if this major part of the plan failed to materialise.

Miraculously, our customer had offered the astonishing sum of £70,000 for the flat, exactly *double* what we had paid for it. We bit their hand off to avoid getting into bridging loans, which Jim Ogilvy had kindly offered us as a safety net. Seventy grand was the key figure in the equation, and the one which meant that my plan had been perfectly conceived, albeit rather ambitious. I felt rich for a nanosecond, and then I was in debt again. Ah well. Needless to say, the £35,000 profit was swallowed up overnight, and it disappeared into the Bermuda Triangle that is our bank account, and our lovely big profit made its way towards the payment of The Old Windmill without even touching the ground.

It was difficult to take in the fact that our place of residence from now on would be this amazing windmill. Although the entry date was ten weeks away, Dad knew the lady in Ryden's shop in Dalkeith (he knew everyone), and this handy acquaintance enabled us to get our hands on the keys any time we liked to survey the place, measure things, and start the process of refurbishment, unofficially, if it was to come to that, as long as the keys were returned to her at the end of each day.

At last we were poised, ready to start work. Although I had always been useless in the DIY department, I was looking forward to taking on the work that lay ahead, and I was willing to try my best, but already I felt that my talents would lie in the *de*construction business, leaving *re*construction to experts. However, I could design things, come up with daft ideas, infuriate workmen with strange requests, and generally co-ordinate all the work. I would willingly become anyone's apprentice, learning a few tricks of the trades, and getting in the way of a whole manner of events.

My Dad's diary recalls the following important dates:

> Friday, March 12 finally came. *I couldn't wait to get away from the American School in Aberdeen, and I skived off early (it was the last day of term anyway and most of the pupils would be doing the same) and I eventually got to Ryden's in Dalkeith at about 3.30 that day. By 4.30, Dad and I were ready for action at the windmill. The first job was to remove the unsightly cupboards upstairs. By the time we arrived in Edgehead, Dad's men were already hard at work. Rab Purves, the brickie, had just started to dig the foundations for the oil storage tank which would soon be situated at the back of the garage to serve the heating system. Dad and I worked till 9.30 on the first night and couldn't wait to get started again the next day.*
>
> Saturday, March 13. *Work started early that day and Lionel Thomas and his mate were on the scene by 7 am (beyond the call of duty if you ask me). When we arrived at 7.30 their tar boiler was already smoking away. The rock asphalt for the small balcony was nearly ready,*

and the balcony was soon stripped and relaid. This was the only job the bank had insisted on being done as a condition of getting a mortgage, which I found bizarre, considering the state of the rest of the place. Despite the obvious problem of water leaking from the balcony into the dining-room below, the men had found all the timber to be in sound condition and soon we were watertight again and the bank was satisfied. The total cost of the first job was £175. Later that morning, Billy Hosie arrived to start the monumental task of installing the central heating system. This was a big job and we knew that Billy would be there for some weeks. His radiators were delivered at 10 am and even their polythene-covered presence improved the look of the house right away. Billy put in a twelve hour shift on the first day, and many more after that, retiring just after 8 pm.

Sunday, March 14. *The brickies returned to build pillars on which the oil tank would eventually be mounted. (This 'working on the Sabbath' is something I'm sure the original builders of the windmill would not have been allowed to indulge in, not until after church at least). Dad asked the men to use their surplus mortar to form a ramp at the entrance to the garage, which they did, and this meant that one day we would be able to drive a car in there without having to jump a ravine. To indicate how quickly extra money, not budgeted for, slipped through our fingers, this wee job cost us £150 and £47 for materials. Indoors, Billy continued his pipe-laying from 8 am. Add it up. It's grim. To soften the blow a bit, we knew we were paying bargain prices through Dad's contacts and favours, and most of the materials came from Knight Bros in Millerhill at cost price, because my Mum worked in their office part-time. Yet it would still cost us an arm and a leg.*

Monday morning, March 15. *By now Billy was ever-present and this would be another long day for him. Luckily we had agreed a price for the complete job, as I dread to think how much it would have cost at an hourly rate with the meter running. On this day, the kitchen was completely gutted, as it was to be the nerve centre of the central heating system. Dad had taken 'holidays' from work and spent the time labouring for Billy. The electricians from John Noble's were supposed to arrive on this day, but they were delayed, and this was something we were just going to have to get used to. Arlene was busy working at STV in Glasgow at the time, and it was not unusual for her to receive a phone call in the studio to ask her to confirm the positioning of the washing machine, for example.*

Tuesday, March 16. *Floorboards were up everywhere and channels had been dug throughout the concrete ground floor with a pneumatic drill. Billy was annoyed because the boiler he had ordered from Trianco had been delayed by a week. At last Noble's electrician started, a young spark called Eric, and the banter was always good between all the different tradesmen. Mum had been phoning round all the fuel companies to get quotes for oil for the tank, and arranged the first delivery for Friday 19th March. Dad went to do some work at night but there were no lights upstairs, and downstairs was swarming with heating engineers.*

Wednesday, March 17. *Billy picked up the huge grey metal fuel tank, ready for connection to the pipes in the house, as the kerosene would be delivered on the Friday.*

Friday, March 19. *With the fuel tank in place, propped on its two wide brick pillars, we were now ready to fill it with 1250 litres of kerosene. For some reason only 1175 litres were delivered and I remember a slight problem with the oil company filling the tank with the wrong type of fuel and holding up the testing of the system, although it was sorted pretty quickly. The cavalry arrived from Aberdeen! Sylvia and Ian (Arlene's Mum and Dad) came to lend a hand.*

Saturday, March 20. *The plasterers, Willie Park and Gibby Oliver arrived to start plastering the big circular bedroom. In the middle of the floor stood a huge blue barrel containing gallons of water, and an orange cement mixer. I had my doubts about how successful the finished article would be, but I needn't have worried. The job was completed at 6.30 pm. Meanwhile, directly underneath, Dad and Ian were painting the big circular living-room. This was a giant leap for mankind, as I really felt a huge difference had been made through one relatively small cosmetic job. The painters also painted the small bedroom upstairs on the same day, all ready for Arlene and me to move into for the foreseeable future. Cut to the kitchen. Billy was tearing his hair out because the boiler which had just arrived (after a week's delay, remember) turned out to be the wrong model. They told him it would be another week before the right one could get here, unless he picked it up himself, from Sheffield. Thank you Trianco.*

Monday, March 22. *Ian and Sylvia had been here for more than a week, and a pile of work had been done in that time. Endless small but important tasks were done all day, some of them dirty jobs that the average mother-in-law would cowk at. Finally, Sylvia returned home to Aberdeen to her work (for a rest) and Ian stayed on to keep Dad company. Lock up your whisky! They finished painting the bedroom today and in the afternoon a pick-up truck arrived to take away a load of rubbish from the garage, because tomorrow we'd need the space.*

Tuesday, March 23. *This was the day the 'flitting' arrived from Aberdeen, although we were anything but ready to receive it, and most of it spent the first year of its new life in the garage which, conveniently, had been emptied the day before.*

It was 'spring break' from the American School where I taught at the time, and this meant two weeks' vacation. However, many months before stumbling across the notion of moving house I had arranged a trip to Rome and Florence with a party of students for the first week of the holiday. I was, of course, guilt-ridden to learn that our only option was to arrange the removal from Aberdeen to Edgehead during that very week. So at the Aberdeen end I was Mr Extremely Unpopular while Arlene and her parents had to pack all our worldly possessions into cardboard boxes and then cart everything on to the back of a lorry. Even as the truck arrived in Edgehead I was still conspicuous by my absence, eating a pizza fungi by the Trevi Fountain while my dear family buckled under the boxes (not to mention the wardrobes, sideboards, armchairs and a grand piano) which were off-loaded, on their arrival, into the garage. And you might like to hazard a guess at who would pay his penance by unpacking the required items virtually single-handedly on his return from Italy.

As the removal truck made its way back towards the Forth Road Bridge at 5.30 that Tuesday night, taking half of the Edgehead treetops with it, Billy's boiler arrived at last, about two weeks behind schedule. Billy immediately got stuck in to the last leg of his job, and we had heat for the first time that night. We all stayed up late in the kitchen just watching the boiler boil, as if it was the first television set in the street, enjoying the warmth and waiting for it to break down any second. All Billy had to do now was to 'set it', and replace a few floorboards around the house where some of the pipes were still exposed, but that was for another day.

Having used most of his contacts from the building trade, Dad had naturally become the foreman. He knew in which order things should be done. Had it been left to me, I'd have probably papered the walls first then realised that they had to be demolished in order to install hot water pipes or something. I was his apprentice, as I had been since I played with my first toy tool kit as a nipper, fixing the grey Morris Minor we had, then the Anglias and a Cortina, not to mention leaky taps, broken washing machines, garage construction, the lot; "no job too difficult, distance no object". I had always been there to hold the spanner, pass the screwdriver or find the pliers or the foot-rule in Dad's oily tool bag, and my DIY skills were at last about to rise up the learning curve under his expert guidance – or so I hoped.

On most days, Arlene worked around the house in an attractive navy blue boiler suit and yellow rubber gloves, and she systematically set about many of the chores which would make the house habitable because, despite it having been turned into a building site overnight, we were going to have to live here at the same time, in the smallest room in the house. The areas not yet to be 'demolished', such as the bathroom and a small bedroom, were scrubbed to within an inch of their very existence by Arlene and the other half of her cleaning team – my Mum (who wore the orange boiler suit and looked like she'd just been Tango'd).

Dad and I wandered the corridors and rooms proprietorially, removing old shelves that belonged no more, lifting old carpets which had been trodden upon once too often, dragging all sorts of debris into the Dalkeith Boys Club Football Team mini-bus which served as a ferry between this new building site at Edgehead and the skips in the Council's yard at Stobhill. At the end of each day, no matter how much mess had been created, Arlene and Mum would tidy obsessively, clean and disinfect the whole place so that you could eat your dinner off any part off the building site. Dad had borrowed an industrial vacuum cleaner which the Clean Team fell in love with, and for a while we all risked being sucked up never to be seen again. Even while under construction, with builders' bums mooning at us from every direction, The Old Windmill was more hygienic than your average hospital wing.

The 'demolition' process was one which involved the removal of things like 'modern' fittings that didn't look too attractive anymore; wood from hatches, cupboards, and shelves were also given the boot as new designs came into play. All of the carpets were for the tip, some thrown from the high windows, others just dragged off to their death, slung into the boot of a car and dumped like a dead body in a gangster movie; and the kitchen had already been set upon by an axe murderer, the only evidence being a winter's supply of kindling and a few nasty splinters. This work was easy and satisfying. It didn't matter if anything was accidentally scratched or scraped or broken because eventually it would all be made to look nice again. We were stripping bare the floors, ceilings and walls to make way for more builders, plasterers, joiners, electricians, and eventually, the decorators.

7 Two Careful Owners

Mrs Clara Chalmers had lived in The Old Windmill since 1962, the only previous owner having been Elizabeth Adair herself, who had had the foresight to rescue its derelict remains in the late 1950s, and whose brother, Hal, was the architect of the extension. Mrs Chalmers, we all agreed, must have been tough, because she had survived thirty winters in Edgehead with no central heating. Even the modern electric heaters *we* used as temporary measures were as effective as a mouth organ in a cathedral, but for Arlene and I the central heating system was an essential commodity. Before finally commissioning the installation, we made sure we had done our research. Gas had been ruled out early on as there was no supply to the village. On reading Elizabeth Adair's book we learned that there was electric under-floor heating still in place – but under a concrete floor! It might have been ultra-modern in the 1950s, but impossible to afford nowadays, as well as being old, ineffective and dangerous. We had Eric the spark disconnect the system on his first day at work.

We knew from the outset that trenches would have to be dug in the floor to lay the pipes for the oil-fired system for which we'd finally opted. The main bedroom, situated on the first floor of the round building, had exposed sandstone walls (very pretty for a dungeon but otherwise cold, damp and unwelcoming). These walls *had* to be plastered up, and we had been offered a multitude of options as to the proper way to do this before making a decision and letting the plasterers loose.

Throughout the house the electrical circuitry was in need of upgrading, as any surveyor would say, and we knew that this was going to be messy. Eric continued to rip out miles of old wiring. New routes were then laid out, and in the end the electrical firm got the award for being the most overpriced group of workers, in my view, but at least everything worked.

In the Secret Room (the room at the top) a complete makeover got underway starting with the demolition of the plywood wardrobe on the first day, and eventually the floorboards were sanded back to their natural condition with an industrial sander borrowed from another of Dad's contacts, Davy Glasgow. The floorboards in the bathroom got the same treatment, but not before the water tank had been replaced. This, after thirty years' sterling service, had decided to take early retirement and spring a leak (perhaps exacerbated by the descaling process we had decided to give it, to clean it up after such a long time), and it being a freak size meant a new tank had to be specially built to fit its awkward space. The versatile Davy Glasgow made the tank too and later a metal security box for the outside tap to deter mindless vandals from deliberately leaving the water running and flooding the whole village. The flat roofs that Arlene's Mum had warned us to avoid received the necessary attention; we settled for a temporary repair until we got some more money together. Like in the twenty-fifth century. The one other more urgent roofing problem had been dealt with in the first week as a priority, and this only involved a few slates and a 'doad' of cement, as I understood it.

Another problem area was the decay of the wood around the sash windows. Thirty years of exposure to the prevailing wind had taken its toll, and during many of these years there was no protection from the tall mature trees which surround the windmill today. One by one, the windows were replaced and the house felt secure.

The bedroom begins to look better.

The warm route down the hall.

Billy's trench warfare with the central heating system.

It took a bright spark to fix this.

The kitchen: a room for improvement.

You've missed a bit: Inglis and Ian transform the living room.

Chesterhill Mill. 1785. The massive sails of timber and canvas turned slowly, the occasional gust of wind being just enough to sustain the impetus. Inside, the heavy mechanism creaked and groaned wearily as it ground the corn from the surrounding primitive farms, brought precariously to the tower by a wobbly donkey-drawn two-wheeled cart.

The mill had been commissioned by an early Earl of Stair some time around 1750, its soft sandstone blocks hewn from an indigenous quarry on the Stair Estate itself. A local millwright had been given the job of building the eminent pile, one of many windmills erected in Scotland during the late eighteenth century and the first half of the nineteenth century, and costing between £200 and £300 each.

As far back as the early seventeenth century King James VI of Scotland conferred by Royal Charter the privilege to build windmills 'for the common and public utility and profit of the burgh'. Through this empowering scheme the landowners were given the right to build corn mills to enable them to grind their own corn only at those mills, and for nearly fifty years this was the sole purpose of Chesterhill Mill. Here it stood like a vast monument; a gargantuan tower, an industrial monster; worked to the bone, giant arms swaying about, round and round, utterly exposed to the harsh climate. Perceptively, in one of his early journals, the well-travelled agricultural pioneer, Dr Samuel Johnston summed up the scene when he wrote of his first visit to the area in the late 1700s:

'A tree might be a show in Scotland as a horse in Venice'.

An English knight concurred when he accompanied the King on a visit to Scotland around the same time, and he wrote:

'Judas had sooner found the grace of repentance than a tree to hang himself on'.

Because of this (well-recorded) absence of trees, Chesterhill Mill enjoyed a perfectly unobscured view on all fronts. Beyond the barren fields to the north east, you could see across the water to Fife on a clear day; then to the south, the rolling hills of the Lammermuirs virtually uninterrupted by towns or villages; and to the north and west, more bleak and untamed fields, unmanned and unmined, their boundaries yet to be marked by protective hedges and dykes. Sickles and scythes were among the few signs of early technology on the farms, and thin livestock wandering over the runrigs were the only suggestion of organised farming life in an agricultural wilderness.

The windmill was designed two hundred and fifty years ago to be blown, and nobody switched the wind off when people decided to live in it. As a result, despite the protection from the trees, much of the outside timber on the premises had to be replaced, and this would happen gradually and systematically as the work went on throughout the summer months.

The exterior of the whole house would in time need attention. Apart from re-harling in places, some pointing to the old part would soon be required, and a coat of paint would be the final cosmetic touch to the extension, preferably in a warmer colour than the current dirty whitewash-white with black and green mouldy streaks running down it. The rusty cast-iron mouldings, which had lost their sparkle, would have to be protected and blackened again for impact.

Then there was the garden; an acre of wilderness, at one time so carefully planned, and meticulously maintained for so long. But after just one year of unavoidable neglect even the Sleeping Beauty's Prince Charming would have jumped back on his horse and ridden off into the sunset.

Soon after we had bought The Old Windmill we learned from the selling agent, who lived in the nearby hamlet of Ford, that Elizabeth Adair had expressed an interest in finding out who the new owners were going to be, and he was able to pass on to us her current telephone number and address. We were equally keen to talk to Miss Adair, because through reading her book, she seemed to me to be an interesting character who had done many things, not least having successfully got this project off the ground at a time when it was much less fashionable to renovate old buildings, and much harder to get mortgages for doing so. Her name was familiar to me through a Scottish country dance tune which had been dedicated to her, one which I used to play on the accordion in a previous life.

On the back cover of her book there was a photograph of the author standing in a snow-covered garden by an old water-pump. She looked smart and sophisticated, and difficult to put an age on. For all we knew the photograph could have been taken in the 1970s when the book was first published, but then again it may have been a photo from 1960 when the original windmill conversion work was being carried out. We played 'guess my age', based on all the information we could find at the time.

Elizabeth Adair was born in Aberdeen and educated at the Aberdeen High School for Girls, which is now occupied by boys *and* girls, and goes under the name of Harlaw Academy. After leaving school, she attended the Royal Academy of Dramatic Art before spending five years on the professional stage in London. At the beginning of the war she returned to Aberdeen and joined the staff of the BBC (where I too had worked for some time) and she became the first female announcer in Scotland. (Since first getting involved with The Old Windmill we have stumbled across many little coincidences all over the place, and they were stacking up at every turn. For example, Arlene's current job *is* as a BBC announcer.) After Elizabeth Adair's broadcasting first, she became a producer at the BBC (done that too) specialising in gardens, animals and the preservation of historic buildings, finally leaving the BBC after twenty-one years to become a freelance journalist and broadcaster. In addition, she has written several books, all about subjects close to her heart. We were still no closer to 'dating' the lady, but we held her in such high regard through her impressive curriculum vitae that the prospect of meeting her was quite scary.

Elizabeth Adair had been inspired to make something of this derelict windmill, when most people would have seen it as just that, and her vision became a reality as a result of her own determination and single-mindedness. Undoubtedly, she had overcome many more obstacles in the fifties and early sixties than we had faced in the nineties. Some time before coming across our windmill, or even the village of Edgehead, she had spotted a passage in a *Country Life* magazine in 1954 which said:

> *'The more unorthodox the house, the more difficult it will be to find a Mortgagee, and the* man *who contemplates converting a windmill on some remote hill-top will certainly have to find the necessary money in some other way'.*

Elizabeth Adair's book had described her first encounter with the building in 1958, her reluctance to buy an orthodox semi-detached within her price range, and her search for something completely different which she could restore. As soon as she saw the windmill her heart was set on it and the rest is history, well documented with the publication of her book in 1977.

We knew that *I Bought A Windmill* was based on a diary of events which Elizabeth Adair had kept since even before the actual work began, and her detailed account of her thoughts, plans, memories and frustrations all existed in her own handwriting in this diary. All costs and purchases are listed, including building materials, trees and plants, and the numerous 'pieces of house' which had been picked up at a variety of sources in between recordings of her natural history radio series 'Birds and Beasts', which she produced at Broadcasting House in Queen Street, Edinburgh. For example, the beautiful turned wood balusters and rails of the staircase in the hall were acquired from an old mansion-house in the fishing village of Fisherrow on the Firth of Forth which had been damaged by fire. Our Georgian doors came from the store rooms of a well-known auctioneer in Haddington and the new wing of the windmill was built around them. There were nine doors in beautiful Baltic Memel pine, brass fittings still attached to most of them, and they had all come from Smeaton House, which had been destroyed in 1950. They cost £5 each at the time and one, miraculously, was concave and corresponded exactly with the arc of the circular living room. All this and more we had learned from the book, with all costs listed in pounds, shillings and pence, as if in a foreign language, or from the Bank of Toyland. We were looking forward to getting our hands on the book's blueprint, the hand-written diary itself.

One day, soon after we moved in, we received a letter from Ian Chalmers (Clara's son) in Kent, asking if we would send him the nameplate from the gate. It was a large heavy iron affair which simply said 'Chalmers' on it. Arlene's Dad was given the task of removing it one morning as part of his duties, and in doing so he brought to light yet another coincidence. Arlene's Dad's full name was Ian *Chalmers* Allan.

(Hum the theme from 'The Twilight Zone' here.)

In his letter, Mr Chalmers also told us that he was going to send us the legendary 'Elizabeth Adair's Diary' which he had acquired from the author, and some weeks later a huge package was left at our front door, too big to fit through the letter box. It nearly went straight in the bucket, as I first thought it was another unsolicited catalogue of fashionwear from Kay's or Grattan's or Next (I was already too fashionable) but I decided to check anyway. On pulling apart the heavily stapled cardboard flaps I discovered a large, old black book with a white label on the front saying 'Tabulated Statement of Touring Accounts'. Inside, the huge white pages were lined across with blue feint lines, barely visible and crossed with superimposed red vertical lined columns, like something from an accountant's office. Each column had its own category pertaining to it.

The categories were:

Town, Theatre, Date, Terms, Gross Receipts, Company's Share, Miscellaneous Receipts, Salaries, Railway Fares and Carriage of Scenery, General Expenses, Authors' Fees, Printing, Total Expenses, Profit, Loss, and Production Accounts.

At the top of each page there were bold headings followed by blank spaces for relevant information. 'Tour of...' would be followed by a space for the name of a production, 'Weeks...' would be for the number of weeks for which the production ran from 1–52, and '190...' was followed by a space for a final digit of the year, suggesting that the book was for use anytime between 1900 and 1909. Scribbled all over the pages, mainly in blue fountain pen, were paragraphs of notes, anecdotes, things to remember, afterthoughts and ideas for what was to become Elizabeth's book years and years later. Additional passages appeared in pencil and some in black ink, and every page was punctuated with exclamation marks, words underlined, arrows, circles and brackets. A red pen then had written bigger headings throughout the diary, and my guess is that they were the final markings made when the author tried to categorise certain pages for easy finding later on. Every paragraph had a date attached to it, and sometimes a precise time. Occasionally there was an angry remark when a passage described work to be done, and the workman had clearly turned up later than expected, or not at all. The pages were alive with Elizabeth Adair's enthusiasm and imagination. They were readable, but it was very hard work. Then in between the last page and the hard back cover, which were Sellotaped together and just intact, I found some old receipts from the late 1950s and early 1960s, presented by the workmen with carefully itemised breakdown of work carried out and cost of materials.

I also found at the back of the book the folded original drawings of the plans of The Old Windmill and its new extension, as drawn by Elizabeth's brother, Hal, a civil engineer, and some letters, one of which was from Elizabeth to Ian Chalmers dated 19th January, 1993 which tells us a bit about the origin of the account book. She wrote:

Dear Mr Chalmers,

Here at last is the original diary of The Old Windmill. I hope you find it of interest – my frustrations and successes! The Account Book it is written in is my father's hand was a record of the touring dates of a theatre company of which he was the manager at the beginning of the century, and the large writing on the left hand pages is his. You will notice the difference in money values between 1958 and 1993. Did you know The Windmill well? – many holidays there with your mother?

With kind regards, yours sincerely,

Elizabeth Adair.

Chesterhill Mill as Elizabeth Adair first saw it in 1958.

The sitting-room: back to basics.

The extension under construction.

Taking shape: the view at the back, June 1960.

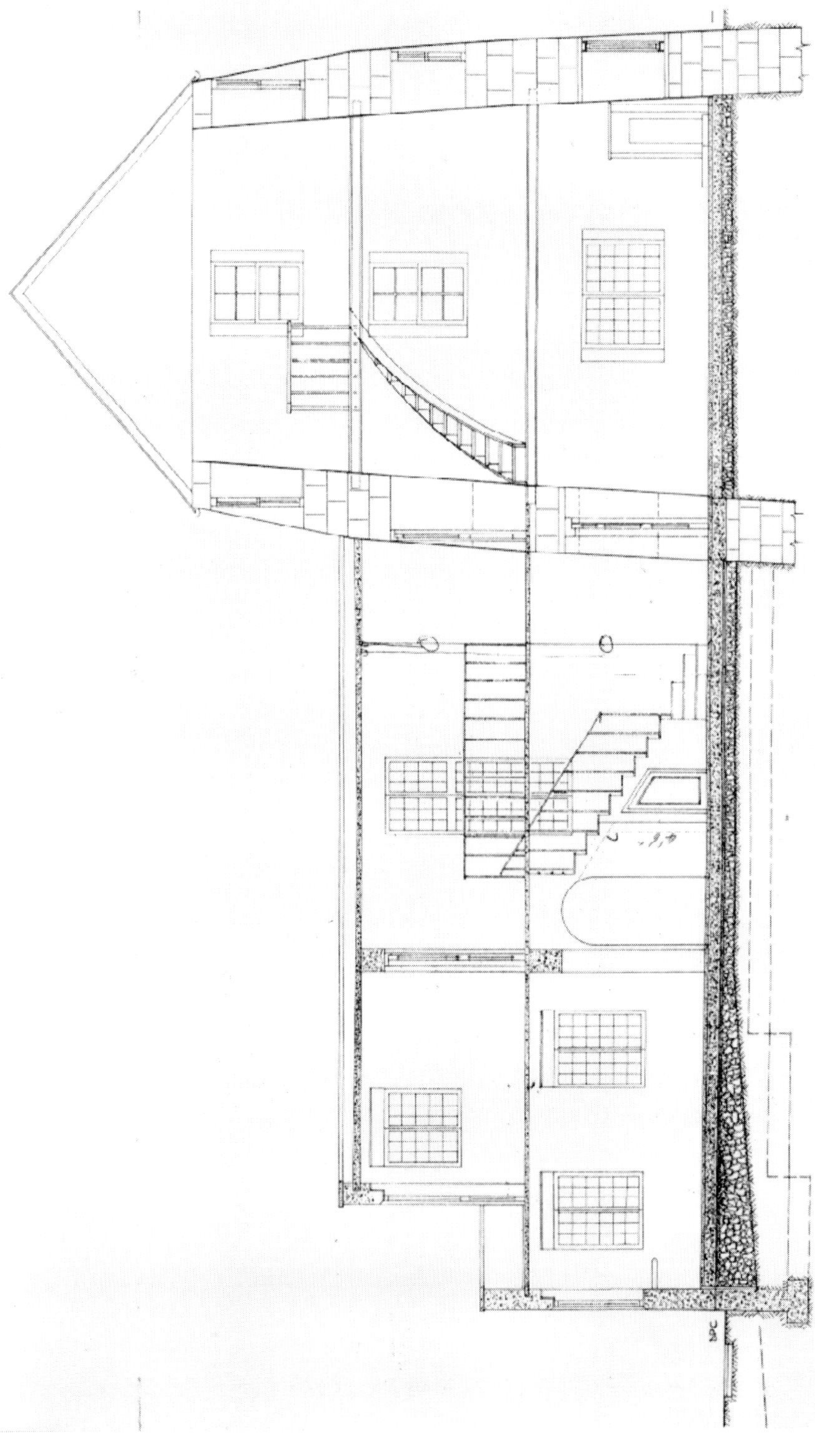

An elevation from the drawings of the Old Windmill prepared for Elizabeth Adair.

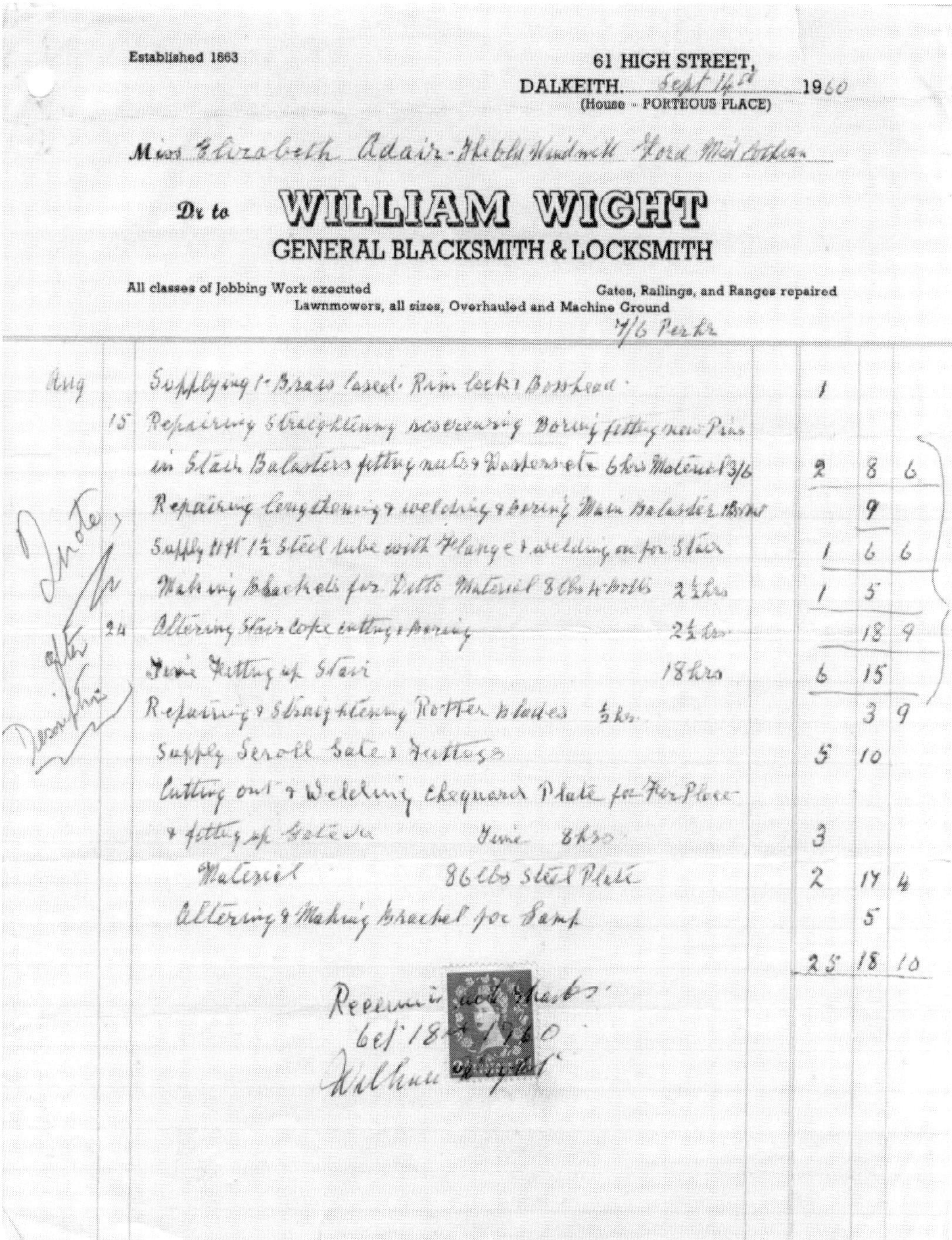

Established 1863

61 HIGH STREET,
DALKEITH. Sept 14th 1960
(House - PORTEOUS PLACE)

Miss Elizabeth Adair - The Old Windmill Ford Midlothian

Dr to **WILLIAM WIGHT**
GENERAL BLACKSMITH & LOCKSMITH

All classes of Jobbing Work executed — Gates, Railings, and Ranges repaired
Lawnmowers, all sizes, Overhauled and Machine Ground

7/6 Perks

			£	s	d
Aug		Supplying 1 Brass cased Rim lock & Bowhead	1		
	15	Repairing Straightening rescrewing Boring fitting new Pins			
		in Stair Balusters fitting nuts & Washers etc 6 lbs Material 3/6	2	8	6
		Repairing lengthening & welding & boring Main Baluster Arms		9	
		Supply 1 1/2 Steel tube with Flange & welding on for Stair	1	6	6
		Making Brackets for Ditto Material 8 lbs 4 bolts 2 1/2 hrs	1	5	
	24	Altering Stair Cope cutting & boring 2 1/2 hrs		18	9
		Time Fitting up Stair 18 hrs	6	15	
		Repairing & Straightening Rotter blades 1/2 hr		3	9
		Supply Scroll Gate & Fittings	5	10	
		Cutting out & Welding Chequered Plate for Fire Place			
		& fitting up Gates Time 8 hrs	3		
		Material 86 lbs Steel Plate	2	17	4
		Altering & Making Bracket for Lamp		5	
			25	18	10

Received with thanks
Oct 18th 1960
William Wight

The blacksmith's receipt.

Sure enough, Elizabeth Adair's father *had* written something in his accounts book, namely a record of when his touring theatre company played Morecambe, Middlesbrough, Manchester, Northampton, Newcastle and Nottingham, and mainly performing what looks like 'Toreador' and 'Orchid', although there are no details of wages, expenses, terms, profits or losses, which is probably why his daughter decided to make use of the book some fifty years later. I hope the taxman doesn't find him. It was a fantastic thrill to open and read this document, even though most of it was illegible unless you concentrated very hard. Dad had the best shot at understanding the mighty scrawl over a period of a few weeks, his glasses perched on the end of his nose and the hundred watt ceiling light on in the dining room, and he took great delight in finding out more about the history of the conversion than any of us knew before. We could see from the diary which material had been developed and used by the author in the resulting book, and we enjoyed other chunks which had been spared publication but were particularly interesting to us.

Using the contact number I was given I phoned Elizabeth Adair at home. She sounded a bit surprised to hear from me, which is understandable as thirty-five years had elapsed since her interest in The Old Windmill was fresh, and to suddenly have a stranger ring her up about this subject must have been odd. A lot of water had gone under the bridge in that time and she had subsequently been active in rescuing other buildings from neglect. For example, after her time at The Windmill she went south to the Cotswolds where her next project was to restore a sad, abandoned little house called Coneygar Lodge. Then there was Cluny Cottage, the Dykehead level crossing house along the Royal Deeside Railway in Aboyne in Aberdeenshire which she bought, saved and renovated in 1966 on returning to Scotland. There was talk of a disused chapel in the Dordogne area of France with a gallery and a small belfry, or was that one in her vivid imagination which constantly looked to the future and embraced new projects? As it turned out, she *was* interested in the new owners and she *had* asked the selling agents to put us in touch, so my phone call turned out not to be too unexpected. Arlene and I were invited for tea the next time we were in Aberdeen.

For now, Elizabeth Adair was living in a cottage called Walkend, in the shadow of Castle Fraser which is in the village of Sauchen near Inverurie in Aberdeenshire, and we were on our way to meet her on a cold, snow-covered afternoon in March. In the Castle grounds there's a long and winding road which leads to a variety of cottages, some occupied by families who work on the estate and others rented by other tenants. After knocking at the wrong door twice (not the same door), we found Walkend Cottage in a pleasant spot with an open view of the estate from all sides. The garden was hidden for the time being in a blanket of snow and the only footprints on the entire estate that day were ours.

We tapped tentatively on the cottage door. A very smartly-dressed elderly lady answered the door, mid-seventies I would say, and she invited us into a warm living room, heated by three bars on full blast from an electric fire. We were made comfortable immediately and offered tea and cakes. On the sideboard was a most noticeable doll's house, like a scaled-down model of an ultramodern home, which had been designed and made by her brother Hal some years previously. It was white, not unlike the new wing of The Windmill and it had two circular towers which seemed to be made from empty food cans. Perhaps one day it had been intended for development in full scale. The tea arrived in proper china cups from a proper pot, making it taste ten times better than tea usually does, and we scoffed cake and shortbread all afternoon.

We talked about many things that day and we deliberately avoided turning the visit into just an enquiry about The Old Windmill, although inevitably we touched on the subject several times. We talked about the BBC, cats, books and music. I seem to remember Elizabeth being quite surprised when we told her of the improvements that were necessary to make the windmill suitable for modern living, and I tried hard not to make it sound as though all her hard work and good ideas had been changed. On the contrary, we admired all that she had done and we were in awe of her remarkable achievements; she was a legendary figure in her field among Scottish women, and now a friend. It was a bit like having Bay City Rollers records when I was nine, and then twenty years down the line I was playing in a band with 'Woody'. That sort of thing.

We got to know each other in the short space of an afternoon, and we invited Elizabeth to come to The Old Windmill any time she wanted to visit, or to stay over. She declined on the grounds that her health wasn't what it used to be, and said she wasn't up to the journey these days, but we were asked to call in again next time we were in the area. Before we left, she gave us a signed copy of *I Bought a Windmill*, which meant I could return the other copy to John Smith at the library.

On a second visit, some months later, the snow had cleared and we were shown the garden, past the garage, in which there stood an old faithful Citroen 2CV, about to be sold. She claimed not to be able for proper gardening any more, although it looked well-cared for to me.

Elizabeth Adair arrives at the windmill.

While we were walking around, my eyes lit up when I noticed the wonderful figure of a small statue called 'The Sower'. I had read in Elizabeth's book that this metal statue of a young man, forever broadcasting grain from his apron, had once overlooked the ink-blot pond in the garden at The Old Windmill. I remembered that photographs and illustrations of him were featured in the book and a story related an attempt to steal him in 1961. Later on he was pictured among the irises in the garden of Cluny Cottage, and I then remembered that Elizabeth had written that he had followed her from garden to garden. After all those years, he was still with her. Suddenly, I had the strong strange feeling that The Sower belonged at The Old Windmill, and I found myself audaciously but politely 'mentioning' to Elizabeth that if he would ever like to come home, we'd be only too delighted to have him back in his original place, next to the pond. Normally I would never have the nerve, but as I say, something possessed me to ask. The reply was that The Sower had been promised to a friend, who would adopt him in the event that Elizabeth should ever move to a house where there is no garden, although I could not imagine that.

We kept in touch regularly since our first meeting. Towards the end of 1994, Mum and Dad were in Aberdeen visiting Arlene's folks, and they too paid Elizabeth Adair a visit at Walkend, spending an enjoyable afternoon in her company. On their departure they were given a copy of her latest book *North East Folk*, on whom she had become something of an expert.

The Sower reflected in the ink-blot pond.

8 Council Business

As early on as February 23, Dad had visited Midlothian District Council's planning department to apply for permission to erect the oil storage tank in the garden. He was told that plans had to be drawn and submitted to the Council, and he had these done on the same day by Rob Laidlaw from next door, in exchange for a bottle of whisky – a bargain. Moving swiftly, he went to the Council offices the next day where he was told several different versions of the planning application procedure, and was led a merry dance, before finally calling on Councillor David Smith, an old neighbour from years ago, to help get things sorted out (spoiling David Smith's night in the *Black Bull*). The following day Dad met Councillor Smith again and everything was sorted out with his help. The planning application cost £51 and the application for a building warrant was a further £20. In addition, we had to advertise in the *Dalkeith Advertiser* the fact that we were planning to erect an oil tank in our garden in the middle of nowhere. Nobody objected. All the appropriate forms and fees were submitted and we awaited the Council's reply with anticipation. They responded by telling us that their next meeting to decide whether planning permission would be granted – and no assurance that it would – may be as late as September of that year. For an oil tank! Well, we laughed. We already knew the required criteria for the construction of the thing, and we had a hunch that permission probably *would* be granted in the end, and if it wasn't, there would be a protest. We also knew that if our work was delayed for six months while waiting for this meeting to take place, we wouldn't be able to move in and all the work planned for the summer would have to be done during the dark miserable messy winter months. No thank you.

With all of this timescale hassle tumbling around in my mind, I took the executive decision that we would go ahead and build the oil tank in secret, and if the Council found out before September and didn't like it, then to hell with them. A week later up it went, built by Rab Purves as per the Council's ridiculous specifications which were outlined on Rob Laidlaw's drawing: two brick pillars, so many millimetres wide, so many milimetres high, in a certain colour, no less than a certain distance away from any building, including a garage, not exceeding a certain speed, blah blah, and so it went on. At last the tank was in place, but it would be the ducking stool for me if the Council found out.

Then one sunny day in June while I was sandpapering a door knob or something, a man from the planning department arrived unexpectedly to deliver notification that permission had been granted to erect an oil tank in our garden, and he should like to inspect the site where the building would take place. I think on that occasion, without having to lie, I led him to believe that I was one of the men working in the house (not untrue) and that I was unable to point him in the direction of the proposed site for the oil tank. I distracted him with a little help of an infinitely more interesting willow tree in the garden (from which Sir Harry Lauder's twisted walking sticks were made, apparently), thus averting his eyes away from the 'scene of the crime' and the monster oil tank which had stood there, half hidden by the garage, fully functional, since the middle of March, and off he went happily whistling 'Keep Right On To The End Of The Road'.

Impatience, and to some extent impulsiveness, on my part meant that the work at The Windmill never stopped, and the house was in a habitable state in a relatively short period of time, even though our daily buzz word had become 'eventually'. To me, and certainly to my Dad, the whole thing was a labour of love. After our part in employing workmen to do most of the big heavy jobs that required specialists, we were more than happy to pay attention to detail and potter at our own pace doing jobs which were far from menial, but saved us a fortune because Dad could do them properly. For many years he had been a painter and decorator, and here was the ultimate challenge in cosmetic enhancement. Within hours of the builders finishing their part in a particular room, we were sanding, skimming, waxing, varnishing, preparing surfaces, and painting and decorating them. In the evenings after work, Dad would be itching to go up to Edgehead and get on with some work. We built a temporary kitchen, added fascias to old wooden cupboard doors around the place, and replaced broken panes of glass. I was using tools I'd never heard of before. We removed all the doors in the house and carted them off to be stripped (because someone had seen fit to paint them in the swingin' sixties) and when they came back we had revealed once more the beautiful wood described by Elizabeth Adair when she first found them, and all they required was more wax and elbow grease.

During the summer of 1993 we were able to get on with a lot of the outside work. We squeezed another year of life out of the exterior window frames, protecting them with special paints and replacing these expensive items only where absolutely essential. The coping stones around the balconies and the roof of the extension were pointed, and we secured the ornate cast-iron decorative mouldings to the surfaces all the way round. We had learned from Elizabeth Adair's book that their star and thistle design was a replica of the trellis surmounting the Scottish state coach which carries the Queen to Westminster Cathedral on the occasion of a Royal Wedding. She had picked up a three foot sample of the ornamental casting from a demolished greenhouse and given it to an iron founder, who reproduced forty-eight copies measuring a hundred and forty-four feet in all. (Not the Queen...Elizabeth Adair). Each section weighed a ton, yet some had come loose over the years, waiting to crown someone someday. Later on, we took stiff wire brushes and cleaned them by hand until the bare metal was ready to be repainted black, as it was originally. One day, at the end of the summer, I overcame my fear of heights temporarily and took black gloss to them all, a fussy and intricate task for the likes of me, and I was as proud at the end of the day as Michelangelo might have been of his average ceiling, apart from looking as if I had just been dipped in the stuff. Dad came home and we admired my work of art from the ground. Then the expert painter kindly pointed out *one* thistle that I'd missed!

One thistle out of thousands didn't glisten in the sun, and I was rumbled by the foreman with the eagle eyes.

A great deal of time was spent in the garden, which miraculously retained a reasonably well-kept look for the size of it. Mum and Dad would do a bit here, a bit there and it didn't look too bad at all. Sylvia and Ian were often down from Aberdeen and between them all they got the place looking more than presentable. Sylvia would often enthuse about interesting plants and shrubs she had found, make up Latin names for them and steal some cuttings for her own garden.

Outside in the drive there was always a load ready for the skip. Old radiators, kitchen units, shelves and endless garden rubbish. We found a ton of loose coal at the back of the garage which Dad and I bagged one night in the autumn, and it burned on the huge open fire in the living-room for the whole of the first winter.

The lesser-spotted workers' tea-break.

From the kitchen window, I watched a magpie fluttering in the copper beech one afternoon. I remembered the opening line of the rhyme 'One for sorrow, Two for joy...' and I scanned the tree for a second magpie, trying to convince myself that I wasn't superstitious in any way. Although superstitions never really haunted me, I had this kind of inner respect for them, as you would with a calm sea, or an unpredictable horse. Perhaps it was a fear of them, or a subconscious belief in them, but I was aware of their existence for a fleeting moment before dismissing them to the back of my mind. Things were going so well, but I never once took it for granted.

Occasionally I found myself trying a bit too hard to disown the 'Lord of the Manor' reputation which I knew some people had perpetrated before even meeting 'the new owners' of The Old Windmill. We needn't have felt guilty. We had all worked hard for this new-found plot, as opposed to having inherited an estate; and there were no spoons of silver in our cutlery polythene bag, from either side of the family. We appreciated our new scene very much and we felt we deserved it. It was all a big thrill, as well as a real achievement for us, and to this day I often have to pinch myself to check that I really am living in a house you would expect a rock star to live in. The Old Windmill soon became our piece of paradise with our own mark on it. Some fantastic place.

But it was more important to me that the family were close together here; long days were spent working hard, evenings on the patio when the light nights came in, meals *al fresco*, and everybody mucking in. My parents and Arlene's parents had clicked right from the start and always got on really well with each other, always comfortable in each other's company. This summer, the first summer, Sylvia and Ian spent a great deal of time at The Windmill, on a 'working holiday' from Aberdeen, just as they had done when we first moved in back in March. They would usually stay at my Mum and Dad's house, and they were as big a part of the operation as anyone; our parents were the kind of people who would never see you stuck, they always bent over backwards to help, took a pride in their work, and knew how to have a good time. Long before us, they had worked out that quality of life was the most important thing in the world. Our parents are the secret of our success.

We had been very fortunate, and I really thought someone was watching over us for a while. All the same, I have been a 'worrier' since I was about seven, and if the truth be told, I didn't much like the look of that lone magpie.

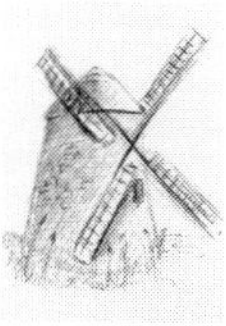

The Shire of Edinburgh, 1812. On the map of this date by James Knox, Chesterhill Mill can clearly be seen on the south side of the road at Edgehead, denoted by the windmill symbol. Prior to that, the windmill's presence on maps is sketchy and ambiguous, although other records confirm its actual existence many years before that. The building, again, makes an appearance on a Map of Midlothian in 1763 without the windmill symbol this time, but it is clearly identified there by the word 'tower' (not to be confused with another tower in a different place on John Laurie's map from the same year).

By the turn of the century (that is from the late 1700s into the early 1800s) improvements had been made to the steam engine, not least by James Watt. All windmills had been carefully situated, by design, in places exposed to the wind, naturally. Despite this, the unreliable winds could not be depended upon to blow to order, and as a result, the windmills began to lose their importance in the corn-milling world.

Rather than face demolition, Chesterhill Mill, like many of its contemporary tower mills, was equipped with a small steam engine and relegated to grinding foodstuffs for cattle. The sails and the roof were subsequently removed and the engine itself was housed in a timber building with a brick chimney stack, adjacent to the tower which already looked derelict and obsolete by this time.

James Knox's map of 1812. You are here!

9 Neighbours and Paintings

Clara Chalmers had been a key figure in the village for thirty years. Within one week we were only too aware of how dear she had been to her neighbours, how active she had been within the community and how hospitable she had been to many visitors to the village. She had been the keenest of gardeners, a loud voice on the village committee, she painted and wrote, and had at one time been an agony aunt on English radio in Hong Kong. By the time we moved in, a bench with a carved tribute to Clara Chalmers and her late husband, Bill, had already been erected in the park by the people of Edgehead, who often stopped at the gate to deliver their fond memories of *dear* Clara.

The windmill had lain empty for almost a year before we bought it and we found that many of the village people were intrigued to find out who the new owners were going to be. It was already clear that we would never live up to the reputation of *dear* Clara and, after learning so much about her, I felt that everyone was in for such a big disappointment. But we were amused to hear the rumours that blew around the village like Chinese whispers. Nosey parkers quizzed workmen for weeks about the character of the new owners. On more than one occasion, I was mistaken for a labourer's apprentice, and interviewed over the fence. Often I would answer the door to be asked if 'the owner' was in, or if I would convey a message to the *lady* or *gentleman* of the house. People in vans selling posh frozen food or tasteless gigantic aerial photographs of the farms in the area would see us as easy targets, or so they thought, and came round flogging their wares. Rather than appear insulted, I left them with the notion that His Lordship would get in touch at His convenience.

Nosiness knew no bounds for one regular dog-walking woman, and I took some pleasure in watching her nose grow during our conversations as I flung her morsels of information, only to hear a fully grown version of the story a week later from another passer-by. This was my new game. Although we were newcomers to the village, I was familiar with the Dalkeith area, and my own undercover network of gossiping aunties would always bring back the word on the street, unbeknown to the village speculators.

'They're awful young.'

'We were expecting a much older couple.'

'He's a concert pianist.'

'She's on the telly.'

'He's on the wireless.'

'There's seven of them living in there.'

'They're from Up North somewhere.'

'They're loaded!'

On the whole though, we were made extremely welcome to the village very quickly, and any trouble was heavily outweighed by kindness and hospitality. I have since played the accordion at the village Burns' Night, helped clean up the woods during Wombling Weekends, baked the flattest scones in the world for Village Day (at the risk of poisoning the entire population of Edgehead) and attended an extremely informative history evening in the village hall. So there. And although the windmill is, in a sense, out of the village itself, more in the woods, it is pleasantly situated for having a blether to the regular passers-by who stop off daily on their way to what must be the biggest dog toilet in Europe. So, you see, we are not hermits, trolls or grumpy recluses after all.

In the early days of living in Edgehead and becoming 'the new neighbours', our lifestyle was somewhat erratic and unsettled. Although our entry date had been March 12, and we were often seen around the place before that (particularly at weekends, during holidays and some evenings), it was some time before we felt we had actually moved in. I was convinced that music teachers in the Edinburgh and Lothian schools were extremely happy in their work and content to close their classroom doors until it was time to come out for retirement, because between 1993 and 1995 not one *open* vacancy appeared for a music teacher. Naturally I had not anticipated this, but I now knew that it had been over-ambitious of me to expect to step into a job in a matter of months.

Instead, I was stuck in Aberdeen in a post that was, it must be said, a fantastic job, but far too far away from where I really wanted to be. I was fortunate, though, not to be wandering the streets or sleeping in my car this time; I was being spoilt during my time in exile by Sylvia and Ian who cooked fine healthy meals every night, gave me wonderful accommodation rent free, and in every way made the frustration of being stuck up north one hundred per cent more pleasant.

At the same time, Arlene was wrestling with the demands of her new job with Scottish Television. TV companies, in our experience, had the habit of lying through their teeth at the negotiating stage, or at the very least moving the goalposts after the deal had been done. For the first few months after our entry date, Arlene spent most of her time in Glasgow, usually away from the glamour of television in a grubby office, often being dumped with editing tasks and scripting interviews with pop stars. At the end of the day she too was fortunate in that she was offered accommodation by my sister, Jacqui, and her boyfriend, Gerry, who lived in the centre of Edinburgh at the time (meal on the table every night and a comfortable place to sleep) but like me, she was only able to visit the windmill occasionally at first. Despite the crazy arrangements that had to be made at the time, it was a fun period, and generally every weekend was spent in Edgehead. Dad held the keys on a day-to-day basis, supervising any work being done, keeping an eye on the place, feeding the cats and getting on with things he enjoyed doing around the house when he got home from work. Every time we returned to the windmill, something important had been fixed; something that we probably would have ignored for years and years if Dad hadn't done it, and when we got home from our far flung corners in the colder months the fire was always set, ready to blaze, so that all we had to do was strike the match.

At times I grew tired of eternally being in Aberdeen, and I often contemplated packing in the teaching lark altogether in pursuit of fame and fortune as a freelance musician. This would mean saying 'no' to a nice regular salary and being paid during endless holidays, and instead I would have to work at nights and weekends, probably playing uninspiring music at least half the time in the worst dives in Edinburgh. It would probably mean saying 'yes' to wearing a frilly shirt and a velvet jacket, and playing sloshes and quicksteps in the smoky function suites of Midlothian at twenty-firsts and fortieths every Friday and Saturday night. I had been there before as a teenager, and the experience had taught me more than most music degree courses at university cover, and at the time I had some great fun and magic moments. But the thought of this becoming my sole purpose in life, the 'occupation' on my passport, and being able to contribute only sketchily to our mortgage made me abandon the idea fairly quickly. My many musical commitments had so far always worked out around my teaching jobs – I had always been able to pick and choose without having to worry about the money, so I decided to hang on just a bit longer before applying for a job as bass guitarist in Danderhall Miners' Club's resident band, who would have been called 'Contrast' or 'Shades of Blue' or 'Face Value' or 'The Last Resort'.

It was wise of me, as it turned out, not to jack in the teaching job, because a fly appeared in the ointment and took us somewhat by surprise. We were already walking a financial tightrope, as most people do when they buy a new house, and our books were balancing to the last penny. The original reason for any upheaval, remember, was principally Arlene's new prospect with Scottish Television, and from the moment we decided it was the thing to go for, our world was turned upside-down and back to front to accommodate the many changes that would ensue. The bank had allowed the mortgage based on Arlene's new salary, her job was to be a springboard to other work and the consequences had a bearing on absolutely everything at the time – family, friends, careers, money, location, even cats.

Nine months into the filming of what Arlene had been told was 'a programme to last for at least ten years', the phone rang very early one morning. I took the call and I could tell immediately that the Scottish Television producer, anxious to speak to Arlene, had his knickers in a twist about something. For a good few minutes Arlene replied only in ums and ohs and there were long pauses in between those. She was clearly trying to take in some surprising, worrying news. With a grim face, Arlene put down the phone and walked over to the window where she peeped through the curtains like a Russian spy, being careful not to be seen. Then she sat down to contemplate what she had just been told.

The programme Arlene presented was a Saturday morning children's show, and one of her co-presenters had been involved in an alleged scandal after the police had raided his house in the early hours of the morning. An arrest had been made and investigations were already underway, and the producer had phoned first of all to put Arlene in the picture, and secondly to warn her not to speak to the press, who it seemed were already on their way to hassle anybody connected with the programme.

In fact, Arlene didn't know anything at all about the circumstances and was as shocked to hear of this as anyone could be. The nature of the alleged crime meant that any association between the accused presenter and Scottish Television was now embarrassing to the company. The case was bound to attract a great deal of media attention over a long period of time and the TV bosses decided that they would disassociate themselves with the now jailed presenter. However, it was decided that the show itself would go on in one form or another. At first the format of the programme was changed completely, in an attempt to keep the bosses 'covered' with as little inconvenience as possible to the schedules, and without drawing too much attention to themselves in the process. Inevitably this meant much more work (no extra pay) for Arlene and the other presenters as Scottish Television clutched at straws to save face while the media circus went mad around them. The programme was doomed. The 'accused' had been such a crucial asset to the style of the programme that without him the winning formula had gone, despite the brave efforts of the others to save the sinking ship. Precisely when it suited Scottish Television, the big chiefs dropped the programme like a hot spud and the whole team was sacked.

In the big scheme of things, we were simply facing a financial hiccup and it was all terribly inconvenient for everybody involved. Eventually, some months later, the producer's number was up too, and even he was given the boot. After the promise of great things by fat cat TV bosses, people putting their trust in the big company and their lives on the line, there appeared to be no loyalty and no thanks, and Arlene joined the dole queue. None of this surprised me in the least and, relatively speaking, we got off lightly. Others didn't fare nearly so well in the broader picture of this particular story, and even after witnessing all the destruction that followed, I managed to remain grateful for what we did have.

The earliest visual record of Chesterhill Mill appears in an oil painting which was done by the great-grandfather of a stonemason by the name of Johnstone who lived in the village of Edgehead in the 1970s. It shows an actual scene from 1848, by which time the mill already had been converted from wind to steam power. The stonemason's own great-grandmother can be seen in the foreground, wearing a blue skirt, sitting on a stool milking a cow. Behind her in its full glory is the windmill, surrounded by small buildings, smoke puffing from the chimneys. The tower in the painting no longer houses any mechanism; no roof; no sails; and its windows are merely slits like in a castle keep. In the distance, a donkey can be seen pulling a cartload of faggots for fuel. The early agricultural pioneers from seventy years before must have made some impact on farming with their humorous quotes, because the picture shows mature birch trees and lush green grass in abundance in the background. The painting is entitled:

The Whun Mill, Edgehead As It Was In 1848.

Johnstone's oil painting of 1848.

From the moment I saw the painting on the cover of Elizabeth Adair's book, I was excited at the prospect that it probably would be hanging in the windmill somewhere, having been passed on from owner to owner; but alas, after reading only a few pages I discovered that the picture was understandably the most precious possession of the stone-mason, his forefather having painted it, and a fond reminder of when he used to play in the mill as a child. Sure enough, no painting. There would be little or no chance of him parting with it, I thought, although I made a mental note that one day I would at least make the effort to cast my eyes on the original canvas, as Elizabeth Adair had been fortunate enough to do all those years ago. Much later, I learned that Elizabeth had offered a good sum of money to the Johnstones for the painting, so that it could be on show in the windmill where she thought it belonged, but they wouldn't sell it. Some years after that, however, a younger generation of the Johnstones came round to agreeing that the picture *should* hang here, and they made a gift of it to Clara Chalmers on her birthday. By the time we set foot in the windmill in 1993, the painting had been removed, and its whereabouts today remain a mystery.

Chesterhill Mill has been immortalised on canvas many times since. In 1947, ninety-nine years after Johnstone's Whun Mill painting had dried, another artist by the name of H.T. Kirk set to work, capturing the windmill in watercolour, showing us the 'new' slate roof intact for the first time. The origin of this picture is unknown and there is no other information to accompany it, but we now have it hanging in the house. The third painting we know of is a watercolour by Clara Chalmers in 1973, featuring the windmill and its extension from through the trees, and we have evidence that it once hung proudly in one of Richard De Marco's exhibitions in a gallery in Edinburgh sometime in the late seventies or early eighties. By chance, I stumbled across another recent painting of the windmill in the Sun Inn near Newtongrange when I stopped off for a drink one Sunday afternoon with a friend, soon after moving into the house. The walls were flanked with paintings by a local artist whose name escapes me, and they were all for sale. I remember not being inspired enough at the time to stump up the £50 they wanted for it, but I did return to the pub with Arlene a couple of weeks later when we were in the area. We went in for a sneaky look, but someone had snapped up the masterpiece and we felt obliged to stay for a fruit juice.

The most recent work to be added to the collection was another oil painting gifted to us at Christmas 1997 by an old schoolfriend called David Sharp, who had left the Air Force after fifteen years' service, and was given an easel and some paints as a going away present (the officers knew this was a hobby David had recently taken up). He told us later that he'd sneaked into the garden on several occasions in the middle of winter while we were out, sat there like Scott of the Antarctic, capturing the tower in the snow.

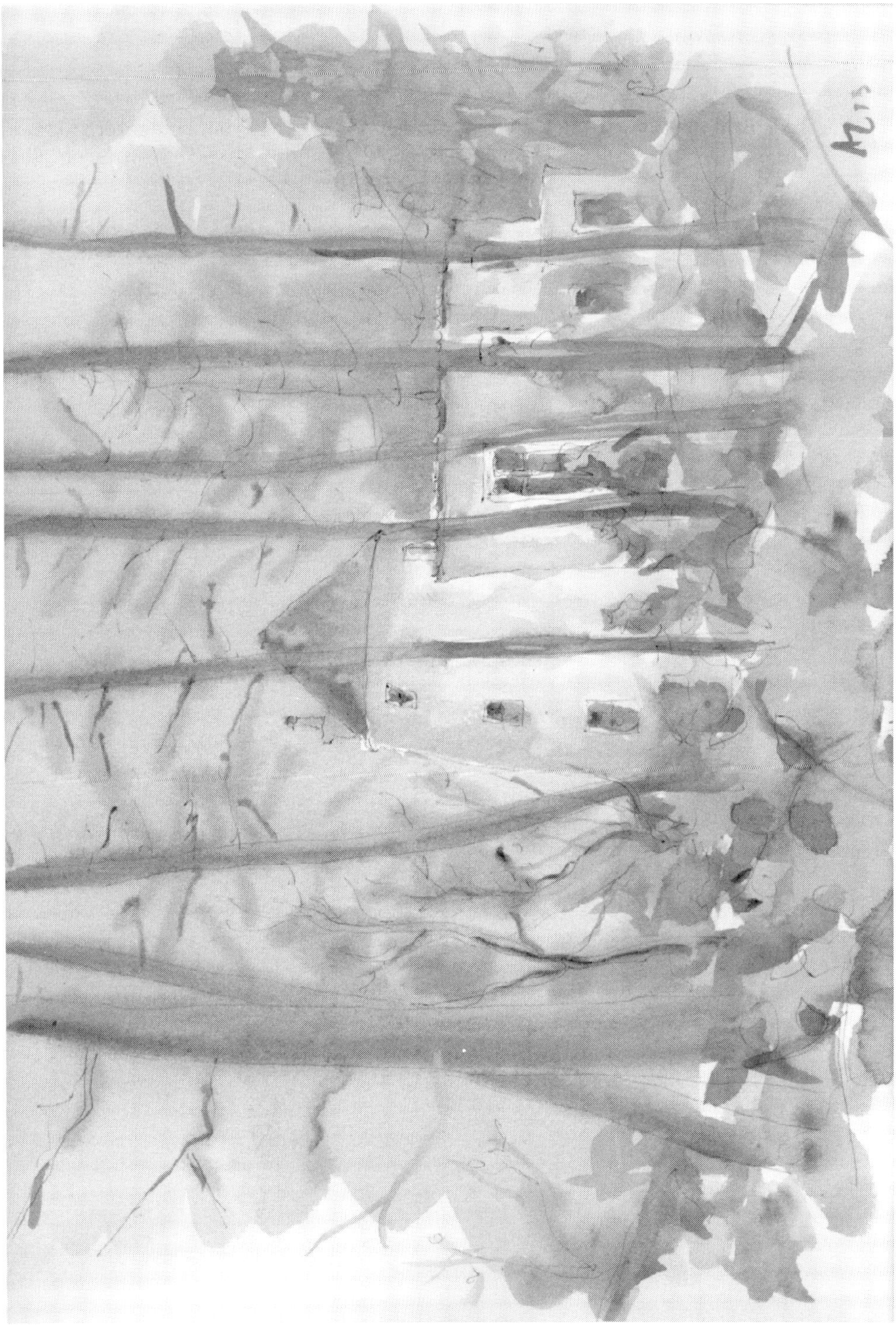

10 Sticks and Stones

One Sunday morning in February 1994, Dad was working like a Trojan, repairing the perimeter fence to the south of the garden, which marked our territory within the dense forest of Windmill Wood. The posts had all but given up the ghost and they were struggling to hold up the wooden slats onto which the timber had been nailed. The whole thing flapped heavily and precariously in the wind and this was a job I would have put off for as long as possible. It certainly wasn't one I would contemplate undertaking in the bleak midwinter. Dad was made of stronger stuff and he insisted that this kind of work was the best thing to get you warmed up on a cold day – his answer to my attempt at reprimanding him for working too hard.

The trees were all but bare bar the pines, the sky was a wintry blue, grey, red and yellow, and the garden and most of Windmill Wood were gripped by the freezing frost that remains in our patch throughout most of the winter months while the rest of the village thaws from time to time at least. It still amazes me that only an extra few feet of altitude can sustain our unique winter wonderland, while even our closest neighbours don't have to do the skaters' waltz around their garden for quite as long as we do each year.

On this particular Sunday morning I awoke to the noises of rural industry, the thud of digging, the hammering and the clattering. I crawled out of bed at about half past nine, and from the bathroom window I watched Dad grafting, as he methodically fixed the fence like a very fast one-man fence-fixing factory. He was wearing a tatty two-tone green woolly jumper that my Granny had knitted for *me* in 1977, and at last it had reached workwear status. A brown scarf tucked inside the jumper kept his neck warm, and the old black trousers he wore had once been his best pair, when piped seams were fashionable in the early eighties. The black casual moccasin style shoes now bore evidence of every single painting job Dad had done in the past five or six years: my sister's new flat, Auntie Win's kitchen, Auntie Margaret's living room, the entire Old Windmill Autumn Collection, and so on. Their colour schemes were all represented in Dad's working shoes, themselves now a work of art. I went downstairs, wrapped up warmly and went outside to keep him company, watching in awe, as usual, his capacity for hard work.

Already the old fence began to look new. In six foot sections at a time he ripped the vertical sarking from the horizontal supports and laid them in sequence on the ground. The rotten supports were then discarded and the flimsy posts were dug out of the earth with a spade, destined for kindling with the other dead wood. Dad would then dig a deeper hole for each new post, painting each one with protective creosote and then dropping it into place, hugging it as if about to toss the caber, before filling in the hole with cement. They were good heavy wooden posts, perfect replacements for the next few years, but length-wise they had not yet been cut to size, so a good extra foot on each new post towered above the old ones which were being systematically replaced. New wood was then nailed across the posts, one strip along the top and one along the bottom, and then the sarking was nailed back into place, only this time more securely. Occasionally Dad would also replace a piece of sarking when the original wood appeared to be in less good condition, masterfully continuing Elizabeth Adair's stockade pattern. Before moving on to the next section of fence, the whole new section was splashed and drowned in more protective creosote.

All this happened at some speed, and a good twenty foot segment had been replaced by lunchtime. The final stage of this job was to cut down to size the new tall posts with a double handled saw, which would have been a handy item to have around for a variety of tasks at the time. We spoke about getting one and I said I would do the sawing off part when I eventually acquired one, but I never got round to it and the tall posts stand lining that section of fence to this day.

More recently, I was thinking about dates and diaries and I discovered that exactly one year before, on February 2nd, Dad and I had taken one of our many prospective buyers' trips to Edgehead and on this particular day we went for a walk down the village. I remember us walking towards The Old Windmill on the way back from the main road and stopping to chat for the first time to the 'next door' neighbours, the Robertsons, who were, well, sunbathing. They were renting the modern bungalow from the owners of the farm as far as we knew, and like every other tenant who occupied the house before and after them, they had made an offer to buy the place. The owners knew what a goldmine they had on their hands and refused to sell the bungalow. Instead, they continue to rent the house to an ever-changing flow of tenants.

I remember the Robertsons' patio doors were wide open and the couple sat on deck chairs in the garden. Toys were strewn across the lawn and Bob had just washed the car. Their nine-year-old son, Ross, lay on a sun lounger holding a pink cold drink in a tall glass with a straw, and he wore nothing but a pair of shorts. From time to time he got up and ran through the water in a plastic paddling pool. We talked for a while and Dad and I agreed the natives were friendly. It was the 2nd of February 1993 and it is a date which has stuck in my mind, because in disbelief at the climate, I repeated it over and over to myself. Such a freak day of February sunshine and warmth has not, to my knowledge, been recorded since. Not in Edgehead, at least.

Dad and I knew of a secret place where a huge mound of pebble stones (or 'chuckies' if you're Aberdonian) had lain for more than a year. It looked like they had been dumped in a school playground to be used as part of a big job, maybe a road repair or something, and they turned out to be surplus to requirement. Nobody bothered to pick them up again and they had no purpose in life, except when children played 'king of the castle' on them every so often. They certainly hadn't been touched for a long time, until over a series of weekends, while taking Dalkeith Boys Club for football training, my Dad began to notice the odd old age pensioner from a nearby 'scheme' filling the odd plastic bag with stones 'for the window box'. The gallas elderly thieves would emerge from the woods, wander around whistling Jimmy Shand tunes while keeping an eye on the janitor's house in the distance, and when the coast was clear, the bag would be filled and the pensioners would calmly walk away, tilting to one side, wee scabby dog trotting around their ankles. Not that Dad was watching too closely, but he soon became aware of how long the gravel had been lying there otherwise untouched. He observed how no work had taken place in over a year, and how the pile was very slowly diminishing, being relocated to the windowboxes of Woodburn. So one day, my Dad decided to join in. He filled a coal sack full of the unwanted stones, lifted them into the back of his car and transported them to the drive of The Old Windmill, which was showing signs of baldness and really beginning to look like the Battle of the Somme on a rainy day. (To me, it was more of a car park than a driveway, given that more than one car could be parked right outside the front door; in fact *twelve* cars could comfortably fit in the 'drive'.)

Carefully spread, the stones stopped us from trailing mud through the house, and they also defined the borders between the muddy garden areas and the drive. By coincidence, the stones were exactly the same washed gravel as the few that remained on the drive from thirty years before, so matching was not a problem.

Admittedly, it took more than one bag to cover the area.

Every so often Dad would drive to the secret location, take a walk, keep his eyes peeled for the janny, and fill the bag up for delivery to Edgehead. As time went on and the redistribution of gravel became a going concern, Dad used to keep a shovel and a stock of bags in the boot of the car, and eventually on a weekly basis, more and more piles of stones would arrive at the windmill. Still, pensioners would occasionally turn up with their small bags or even wheelbarrows, and the pile would disappear by degrees. The change was not really visible to the human eye, so gradual was the erosion of the stones, but some months down the line, it suddenly became more obvious to us that the mountain of unwanted chuckies was a shadow of its former self, while up at the windmill we had a beautifully covered gravel drive. Some time later, when the programme was nearing completion, I would go with Dad to help load the last few kilos of stones that we required from the pile into the car. Sometimes we would drive in past the janny's house on the way in, give him a friendly nod, park the car, and go for a walk on the huge playing fields beside where the stones were, and maybe take in a bit of a football match on the nearby pitches. Half an hour later we would drive back out, sweating like pigs, with the rear end of the car scraping along the ground and the front of the car in the air as we left the premises. Polite wave to the janny on the way out, thank you very much and off we went. Hardly the Great Train Robbery of 1963, but lots of fun, your honour.

The tons of stones served us well for at least a couple of years before the drive had to be reinforced even more, but to this day, at a secret location, somewhere on the outskirts of Dalkeith, not far from here, there remains a considerable mound of pebbles, unused, unclaimed and unwanted, and ripe for the picking.

The drive in 1962…

… and then in 1993, before the stones arrived.

11 One Year On

Our first anniversary was almost upon us and within the year we had transformed The Old Windmill into a very comfortable dwelling place, and although the work was on-going, virtually all major repairs had been completed. Wandering around the house, it was easy to forget how far we'd come and how much we had actually done. I thought back to some of the jobs that made all the difference.

In the huge round living room we'd had the chimney swept because we were never sure how effective the fire would be. It looked superb and the Adam fireplace was the ideal surround, but there was no telling whether or not it would work. The sweep checked it for dead birds and other obstacles and used smoke pellets for further research. Eventually he confirmed it had a good draw, and sure enough it became the perfect centrepiece, but not before a new slate hearth was fitted to set it off. Two radiators were now in place and the trenches which led the pipes to them had finally been cemented over, smoothing out the concrete floor. Under there, somewhere, lay the underfloor electric heating system which we were unable to actually locate. I was assured by Eric that its cables had been disconnected though, and that no trace existed, but I shuddered at the possibility of an accidental reconnection and how much the quarterly bill would be. It was a potential Frankenstein. Cosmetically, there had been very little to do in the living room, except fill some cracks and apply the paint – a warmer colour than the stark white which it replaced. I removed two ornate wall lights which were fixed above the fireplace, and twelve months on, I regretted this hasty decision. The iron chandelier which hung in the centre of the room had to be rewired after some testing, and the finishing touch was a repaint in matt black, which I watched Dad do with a very small brush and a keen eye, without having to remove it from the ceiling.

Upstairs in the equally huge and equally round bedroom, there had been an absolute transformation. The grim interior sandstone blocks had been plastered successfully and now had a thick smooth surface, which was bright in comparison to the dark stone, although it took many days and a lot of fresh air for the plaster to gradually dry from dark grey to white. This was another tradesman's masterpiece and not an everyday task for our two plasterers, who had arms as long as gorillas' by the end of it. In this room another two radiators were fitted and floorboards had been ripped up to make way for their central heating pipes. In time, the ancient wooden boards were put back, and now a path of heat flows from my side of the bed across the room, through the bedroom doorway, right along the hall and into the bathroom, following the route of the pipes, as if by design! No need for slippers, but the cats have also discovered this 'equator' and are regularly kicked out of the way if I am to avoid a cold footstep in the middle of the night.

Elizabeth Adair's glass-fronted bedroom door with the piano music stand design was removed in the summer and replaced by our own design of old French shutters from another ruined stately home, which were converted into double doors as an entrance to the bedroom. Yes, it was as crass as it sounds and they, too, were sacked and carried off to the garage after only a week in service. We were unable to get our money back because we had chopped them up a bit in an attempt to make them look right, which they never did, and now they were too deformed to be either doors or shutters. After kicking ourselves for wasting money on poor ideas, we finally replaced them with a beautiful old door which had to be cut down a bit to fit the space, but it's much more in keeping with the house. An impressive ornate mirror which had come from Rosslyn Chapel once had enhanced the bedroom's look, and after reading Elizabeth's book I had looked forward to seeing it. Unfortunately, it had been replaced by a balsa-wood fitted wardrobe at some point in history, and this obviously got the chop when I got round to it, revealing that it had been built into the site of the *original* chimneystack. This was an interesting discovery, confirmed by the fact that the cupboard upstairs in the Secret Room could also be seen to be an ex-fireplace, and even downstairs in the living room there was an alcove, not designed to house my piano (its new role in life), but once the site of a third fireplace. When and why the fireplaces at one side of the circular house were decommissioned and another one for ground floor use only opened up, we have not been able to establish. Further evidence of chimney changes can be seen on the exterior of the building, as if huge blocks of stone had been reconstructed many years ago. (Later, a final episode in the chimney saga saw us replacing the chimney stack itself after too much wear and tear, but this turned out to be quite straightforward). So the balsa-wood fitted wardrobe was removed and a gaping hole was left looking like a liftshaft with no lift.

Edgehead doesn't have a shop in the village. It doesn't have a pub (although it boasted five in the 1950s, presumably to serve the thirsty mining community which no longer exists) and it lacks even a Post Office. But it does have a joiner's yard, and that is where we went to employ the services of the man who would cover up the hole with an excellent fitted wardrobe. A poor substitute for the piece from Rosslyn Chapel I felt, but somewhere for Arlene to hang her clothes at last, and the new cupboard was built in a day.

One striking feature of the round bedroom is the spiral staircase which winds its way up to the Secret Room. We read a detailed account of its origin, its complicated construction and its cost in guineas in Elizabeth Adair's book, and all we had to do was simply paint it, which was a fiddly enough job for me. I did the first coat of fresh white and took forever about it, then Dad gave it another two coats in less than half the time. He then went on to paint the white ceiling in the bedroom and colour in the walls (or wall). Arlene chose the colour scheme for this room and I expressed my doubts about the peach, the wild flowery curtains and the indescribable carpet, but once everything was in place it looked great and I ate my hat again.

Stairway to heaven (photograph courtesy Paul Graham).

When the paint had dried on the spiral staircase we were able to complete the work in the room up above. The horrendous wardrobes had been chucked out on day one, as had the rust-coloured carpet since it no longer fitted the full perimeter of the Secret Room. Again, floorboards had been ripped up so that pipes for the central heating system could be laid, and I wondered how effective any heating system was going to be in a high-ceilinged room at the top, inside the giant cone. Twin radiators were given the task of keeping the place warm, and only time would tell if they would do the trick. We had used Davy Glasgow's industrial sander to strip the white paint off the floorboards to reveal beautiful clean pine, which we smoothed and varnished. The walls retained their natural stony contours and although they were in good condition they were in need of a fresh lick of paint. Arlene chose the colour again – warmer than the existing white – and she, Dad and I set to work dabbing gallons of emulsion into the somewhat absorbent sandstone walls. Some hours later we stood back and admired our spectacular achievement. The room was cosier, warmer, cleaner and bigger. The bad news was that Arlene announced that the shade should be *slightly darker*, and who were we to argue with Mein Kommandant. Without any debate or hesitation Dad mixed the whiter shade of pale with a darker shade in the pail and within seconds six arms and six very large paintbrushes were again dabbing the stones until the *slightly darker* colour had covered the area. We seemed to work faster second time round and at the end of it we all stood in the centre of the circle looking outward at the walls, Dad and I awaiting a thumbs up or a thumbs down from Empress Arlene. Luckily she was delighted with her new choice and we weren't sent to the Tower. Soon after, on a sunny summer's day with the sunlight from the blue sky pouring into the Secret Room, I enhanced the colour of the walls by painting the small built-in cupboards which remained from the late 1950s. Dad had the more intricate task of painting round the four sash windows, inside and out, and the upstairs job was complete when the warm wind had dried all the paint. The whole tower part of the house was now more or less finished, although there would always be something to do, to keep ahead of the game.

One of the first items of furniture to be taken from the garage into the house was a beautiful old oak teacher's desk from around the early 1950s. We had bought it for £120 from an eccentric wood restorer in Aberdeen who slept in an old corrogated Renault van in his wood yard in Jopp's Lane. He was, I recall, an ex-lecturer in marine biology at Aberdeen University and he had a passion for antique pianos, but he had 'gone mad' and ultimately lost his job. He was often sighted cycling round Aberdeen's big Mounthooly roundabout the wrong way, round and round, causing untold chaos. Desperate for cash, he set up a wood restoring business, and when we turned up at the yard looking for 'an old teacher's desk', he took us to his flat in University Road in the College Bounds, where the very thing stood under a pile of rubbish. We could see why he chose not to sleep there. The flat was filthy beyond description and unfit for even pigs to live in. Two items of furniture occupied the stinking flat. One was a very early full size grand piano from the 1700s, which took up two rooms and must have been worth a fortune. The other was the old teacher's desk. It was exactly what we were looking for and we took it away, holding our breath most of the time because of the vile stench. I spent a lot of time cleaning and restoring the desk, unaware that many years later it would be the perfect piece to enhance our Secret Room.

One night after working all day, Arlene and I decided to carry the desk upstairs before going to bed. When we got to the spiral staircase we realised that our judgement was haywire and the desk would never go up the spiral, so we went to work with a screwdriver, dismantling only the side panel first, so that the desk would then fit up the stairs easily. Or so we thought. It seemed we had to remove the other side panel too. That didn't help much either, so we removed two of the legs. Sweary words filled the night air. Remove the inner support beam. No use. Unscrew these twelve screws. No help. This went on for hours. Time after time, we were convinced that only one more screw had to come out or one more leg had had to come off before the desk would fit through the hatch. This was the epitome of a square peg in a round hole scenario. Eventually, at three o'clock in the morning, we ended up carrying upstairs a bag of sticks, four legs and five drawers, and then we set about reassembling the bloody thing. I suspect the desk will remain in the Secret Room forever.

It became difficult to picture The Old Windmill in the sad state in which we had found it. After only a year I continued to take stock of the changes we had made. In the extension to the windmill, we had decided that the enormous oversized hall (where another bedroom could just about have been made) and the adjoining dining room would become one large entity as far as the decor was concerned, with similar colours and carpets. The walls were in need of wall covering, as a lick of plaster could not have saved them from looking too craggy, and although I liked the old colours very much (unchanged since the sixties) they were in desperate need of attention. The house smelled of emptiness, dampness and neglect when we bought it, and this reason alone was enough to compel us to freshen up the whole place with paint.

Significant rewiring took place in the hall, partly due to dangerous existing wires, partly due to wall lights I wanted ripped out, and partly due to re-routing cables for new ideas we had, not least the upgrading of the burglar alarm. By the time these tasks had been done, as well as the installation of another dozen radiators, the only way to cover up the mess was with wallpaper, and so Victorian-style embossed paper was chosen for the ceilings in the two levels of the hall and the dining room, and a lightly textured design was picked for all of the walls in these areas. Arlene went to Strathearn Paints, where Dad got substantial discount, and chose a wallpaper design that wouldn't jump off the stairs and give you a fright.

There had always been minor problems, drawbacks and unforeseen circumstances in our way. Many of these were to do with the co-ordination of workmen. If one man didn't turn up on his appointed day, others would have to be postponed because they couldn't get in to do their job until the first man's job had been done. Sometimes when particular materials had been ordered, the wrong items arrived (Billy's boiler for example) and this, too, held up the work. It was very frustrating for me because although I was on site most of the time, I knew I was too useless to tackle any of the important tasks, and Dad usually ended up making phone calls on my behalf to make the workmen get their act together when necessary.

On one occasion I remember Dad and Ian were papering the ceilings of the 'big' area, the hall and the dining room, with the aforementioned Victorian-style embossed paper. The paper was absolutely fine, and the Dads had nearly eighty years' experience of papering ceilings between them, but the walls of The Old Windmill and its wonderful extension are not exactly straight as a die, and the job became a bit of a headache for both of them. There were subtle alignment problems with the walls and real inconsistencies on the ceilings, and it took all their expertise and patience to get it right. Lesser men would have been happy to fling up the paper and make do, but these men took a pride in their work, and had been nagged for thirty years each by carping wives, so there was no room for error or imperfection. It never even crossed their minds that they would be beaten by a DIY difficulty. And sure enough, they were not. The result was beautiful and unflawed. They had somehow managed to make both paper and walls appear straight (an optical illusion to this day) and the final effect was perfection itself, albeit after a lot of blood, sweat, tears and expletives.

The two Dads deservedly had the rest of the afternoon off, enjoying the sunshine, walking in the garden, maybe pulling the odd weed. They took a well-earned rest from the day's toil and were really quite pleased with their achievement. It was also Billy the heating engineer's last day. I remember he arrived late that afternoon to set the boiler and replace the last skirting boards and floor boards. Billy had been dropped off at the windmill by his wife, Julie, who wanted to borrow his car and therefore she would have to pick him up again around nine o'clock that night. The sun was going down around half-past nine, and Julie returned with their two-year-old daughter in tow. They waited a while as Billy worked on, and the wee girl ran around the house as terrible twos do. She ran in and out of the kitchen, round and round the living room like a motorbike on the Wall of Death, and up the stairs and along the corridor towards the round bedroom. Julie chased after her daughter, the clatter of high heels on wooden floor boards hammering through the hall. As I stood on the ground floor looking up admiring the newly dried 'wallpaper masterpiece' that Dad and Ian had slaved over all day, a woman's stiletto shoe, with a foot in it, appeared through the ceiling, and wood, plaster, and bits of embossed Victorian-style wallpaper and industrial fine mesh rained down on me like a scene from Laurel and Hardy. That's another fine mesh you've got me into, I thought. Upstairs, in all the chaos and the muddle of the renovation's uneven terrain, Julie hadn't realised that she'd put her foot in it. Billy was in the kitchen all this time, and he and his family were blissfully unaware of the tragedy, and I didn't have the heart to mention it, as he had grafted quite reliably on our behalf at a fair price. In addition, I liked Billy and his family a lot, and he'd even lent us his caravan to go to Arran that summer when we couldn't afford a holiday, so I said nothing. After I had waved them off the premises with a fake smile on my face, like when you pose to have your photograph taken, I scurried upstairs in a futile attempt to repair the damage, but it was no use. That night I watched two grown men cry into their whisky.

The dining room had suffered from dampness, and soggy plaster was peeling off the walls when we originally viewed the house. The problem lay in the balcony of the small bedroom above, where a leak in the coping stones had caused so much damage. Our budget had only stretched to a temporary repair to satisfy the bank, and this repair had been one of the first undertaken. On experiencing our first Edgehead winter, we knew that eventually the job would have to be done again more thoroughly. We couldn't complain because the job did hold out for four years, eventually showing only light stains on the ceiling and the return of slightly soggy plaster if you got up on a chair and searched behind the curtains for it. Dad had papered the dining room single-handedly, and had replaced a lot of the rotten timber on seven of its leaded windows. Any more glass and this room would have been a conservatory. He continued to work flat out around the house and finally I managed to persuade him not to paper the high walls in the hall himself, but instead to let me employ painters and decorators to help just with this one project. He agreed, and recommended a father and son team from Gorebridge, both called Charlie Falconer.

The two Charlies were employed for three days in June and did sterling work. Charlie Senior was in his seventies and addicted to tea. After only one day I couldn't keep him in the stuff, so I went to a Mace shop where I bought a box of bargain basement tea at sixty pence for a hundred and fifty teabags. With the number of tea-drinking workmen coming and going, this could be a shrewd financial move, I thought. But after one cup of the cheap tea, the two Charlies politely refused any further offers. Instead they turned up the next day with about a week's supply of *drinkable* tea in a variety of thermos flasks. Most of the other workmen shared the Charlies' taste in tea and I was left with a box of dodgy dishwater teabags that nobody else would touch.

There was little I could do around the house while the Charlies were at work in the halls, so I decided to mow the lawn for the first time. Sylvia and Ian had bought me a petrol-driven lawn mower for my birthday. Just what I always wanted. I would have preferred a bass guitar or a motorbike, mind you, but in retrospect we couldn't live here surrounded by nearly an acre of grass without a mower. It was a bright, pleasant day with a bit of a breeze and the fresh air would probably do me good. So far I had put off any garden work, simply because I didn't think it was a priority, but luckily others disagreed with me otherwise it would have become overgrown to the point of no return. The parents would regularly sneak in a bit of hoeing here and a bit of pruning there, just managing to keep it under control while I was busy making other plans, usually stupid music industry pursuits, and so with the arrival of my new machine and grass up to my knees, I had little excuse to sit on my backside.

The lawn mower, I must point out, is not one that you sit on like a tractor, but it does have an engine and a clutch and you run around after it as it devours your garden like the Tasmanian Devil. Before starting her up (by means of a pull-cord which takes no less than twenty pulls to start, instantly putting you in a bad mood for the rest of the day), I wandered the sprawling lawns, trying to decide where to strategically dump the grass so I could plan my cutting route and my drop-off points. Soon after, on my maiden voyage, on the way to one of my newly appointed dumping tips carrying a ton of grass, sweating like an Arab's sandal, and gasping for a cold drink, I looked up to the large balcony above the front door of my house to see the two Charlies sitting stripped to the waist, soaking up the sun, sipping mugs of Tetley tea and taking in the view of Soutra, as they waved down at me like the French monarchy. And I thought to myself, 'what's wrong with this picture?'

Arlene's Mum, more than anyone else, was staggered at the rate at which we had made The Old Windmill habitable for the late twentieth century. I knew she would have preferred us to take our time about getting the work done, and would have been quite happy to see the completion of one room every six months. She often referred to my impatience, haste and impulsiveness, but I'm sure she was really impressed that we covered so much ground in so little time.

Before Elizabeth Adair left in 1962.

After the Hutchisons redecorated in 1994 (photograph courtesy Paul Graham).

The dining room in 1962…

… and over thirty years on.

The downstairs cludgy as found in 1993.

The last room to receive a makeover two years later.

Elizabeth Adair's original bathroom with the European hip bath in 1962.

The decor changes, but the panoramic view from the throne remains the same.

12 Seasons Change

I looked out from the balcony to see the day's washing blowing in the wind like a contemporary dance group at the Edinburgh Festival. This was another one of our wee, almost-forgotten jobs. Early on, Dad and I had dug a hole outside the dining room and poured cement into it so that the whirly drying contraption could be planted in the garden, and there it was, still standing a year later, to my amazement. We had come a long way since planting a whirly.

We have the most pleasant view from any bathroom I have been in. We had decided to keep the hip bath as it was in perfect condition and such a novelty, and I have never seen another since. Tilers were employed to cover part of the room in large tiles that suggested an aqua theme, just after the new water tank had been installed. Dad and I used a thing called a 'mitre box', if my memory serves, to make mouldings to enhance the huge cupboard doors which until then had looked plain, and new wood around the leaded windows was installed in this room too. We had sanded the floorboards and begun to stain them. Dad completed this job while we were on holiday in Arran, varnishing the beautiful dark wood floor for protection from the new shower, and applying some striking blue paint to the walls. I don't know if the work was running out at last, but I remember three of us: Dad, Ian and myself fixing a toilet roll holder to the wall in here.

Ian and Dad had also given the small bedroom a fresh lick of magnolia to brighten it up temporarily, and the work in this room was more or less complete. The house was soon ready for carpets and curtains, which had already been chosen during painstaking visits to a quality house furnisher in Inverurie, and miraculously we were on schedule to receive them at the agreed time. Despite the quality of the items, carpet suppliers are incapable of keeping to an agreement, and five years on we await the arrival of the rest of the order.

Only the kitchen and the small downstairs toilet (which you couldn't get into for wall-to-wall, never-to-be-unpacked boxes) were to be dealt with as part of the next phase, some other time.

Arlene and I had spent the first winter mainly paying attention to detail, picking up knick-knacks and bric-a-brac, or 'junk' that Arlene's parents said they wouldn't give house-room, and plans to tackle the exterior were put in place for the spring and summer of the following year.

Dad and I paced out measurements for plans to build a further extension someday, exchanging ideas as we took giant steps from the back of the house almost as far as the perimeter fence, and our imagination went to work in detail fairly early on. In reality though, lack of finance would prevent this from happening for some time, but eventually one day we might be able to make use of those early plans, and I shall have a place to house my jukebox and my drum kit.

We had decided to paint the outside of the existing extension a warmer colour than the shabby stark white which was now looking quite drab and scruffy. After weeks of discussions and browsing through colour cards, Arlene agreed on a colour called 'sand' which was not unlike the famous stately 'Inveresk Orange' but perhaps less shocking. Dad had taken a week's leave at the end of July to do the job and I was on my eternal school holidays, so I was there to help him. Before we set to work we realised that the harling was in need of replacement in certain areas, mainly above the front door and below the west balcony, and to paint over loose harling would cause big problems later. The not-quite-so-loose harling had to be chipped away from the brick using special chisels and I remember this being the pig of all jobs. I was utterly useless at hacking the stuff off (although it certainly managed to hack me off) and I remember leaving it one day for Dad to tackle when he got home from work after I'd given up. In addition, we employed a man, to do the re-harling, who was less than helpful. He actually refused to chip off the old stuff himself, but came back to do the easy bit. He must have known what lay ahead, the swine. What's the use of having a harling business when you refuse to harl? I never forgave this man for giving us poor advice. When this hellish toil was finally over, another formidable assignment faced us – the painting of the exterior.

Having said that, we were now experts in this field and we set to work stirring the gallons and gallons of paint, ready to transform the look of The Old Windmill. Much of my time was spent securing the bottom of ladders with my splayed feet while Dad was up there somewhere with a roller on a giant extended pole in one hand, and a tray with a supply of paint in the other. Even then he made endless trips up and down the ladders for more paint. He would balance on one foot on the top rung, stretching precariously to reach the far corners of the building. I got to gloss drainpipes, refill rollers and fetch drinks of cold water. Sometimes I managed to paint sections of wall closer to the ground. It had taken a day or two to do the preparation before painting, including the masking of all the windows with tape and newspaper. From a distance, the windows had the appearance of colourful stained glass while the job was in progress, but close up you could see that they had been masked with the colour section of the *Daily Record*. It took a full day to apply one coat to the extension, but even after the first coat the house looked like a million dollars, compared to its appearance only the day before.

The weather had been good to us for this kind of labour. It was warm enough for tackling outside work but not uncomfortably scorching hot, as July days tend to be every ten years or so in Scotland. Also, there appeared to be no threat of rain and there was only a soft wind. We worked to the music of Rodriguez's Guitar Concerto and Nat King Cole, among other CDs which I had blasting from the house, the big hi-fi speakers turned towards wide open windows, providing a pleasant soundtrack to the summer.

Read all about it: new colour for old windmill.

The final touch to the exterior paint job was to paint the window surrounds to sharpen up the new colour. We gave Arlene two choices of colour to match the 'sand'; one was pure clean white, and the was other a creamy off-white which Dad and I preferred. We painted some window surrounds in the white and some in the cream so she could see the two versions in context, and we waited for her to choose. The pure white got the autocratic vote and Dad and I silently slipped into our overalls again and painted out all the cream, then walked round admiring our latest handiwork. In the ensuing days, many passers-by on their daily dog walk, who were so used to seeing the old building deteriorate slowly, would stop to compliment us on our work, for this was a real improvement, it has to be said. As we were tearing off the last pages of the *Daily Record* from the windows, a woman who said she had come from North Berwick stopped on the private road (which she didn't officially have permission to walk on) for a good long look. There was no doubt that she was admiring the new-look windmill, but then she couldn't resist bringing a kind of doom and gloom to the happy finale when she leaned over the fence and whispered in a loud menacing voice,

'Did you get planning permission to change the colour?'

There was an awkward shuffling of overalls and paint-brushes, and I cunningly bought time to contemplate her question by patting her scabby smelly drooling hound on the head through the bars of the front gate, pretending to like the thing. (Both the mongrel and its owner had probably ignored the Council's 'No Dog Fouling, Penalty £50' sign outside our patch at the edge of the wood for years, but that was a cheap shot and I resisted mentioning it.) There came no quick and direct answer from either of us as one woman and her dog waited for a reply. Dad and I looked at each other cagily, and then *he* slipped off round the side of the house, diplomatically pretending not to hear the question, and I lied to her through my teeth.

From the day the outside of the house changed colour at the end of the summer of 1994, I really felt that the work was close to completion. The Old Windmill now had our stamp on it. It was clean, fresh, habitable and everything worked. It was restored, improved, stylish and homely, and it was everything I had wished for and dreamed of, as far as buildings were concerned anyway. Sure, work would be ongoing for some time to come; a modern kitchen was lacking, the downstairs bog was still stuffed full of unopened boxes, and dozens of little jobs would be done at our leisure; but on the whole, it was a *fait accompli*.

The first of our dozens of little jobs, as it turned out, would be to prepare the wee bedroom upstairs for an addition to the family. The stork was coming in February or March and this would be a new and happy experience, most people told me. I was excited and terrified. Within a week of the news breaking, Mum was knitting a tiny woollen garment, complete with hood and buttons, that Dad immediately christened 'the duffel coat', although it looked like it was for the Grand Wizard of the Ku Klux Klan, and suddenly we were all preparing for something big. We all had a new focus, something that none of us, I think, could ever imagine happening, but before long another wee Hutch would become a reality. Eek.

Soon the leaves on the trees, too, had changed colour and The Old Windmill looked splendid among the autumn shades, as if we had carefully and deliberately chosen the tone to look its best during this very season, against its new golden backdrop.

A month later, in September of 1994, Dad went to the hospital for some tests. They found what the consultant described as 'a little bit of mischief', but things eventually began to look bad. By Christmas, his health had gradually deteriorated but he remained the life and soul of the party in every way, still insisting on doing light jobs, staying active, doing the things he enjoyed, and particularly looking forward to the birth of his first grandchild so very much. He did everything possible to keep going and to keep living. He never once complained to us about his illness and we, the family, still came first as we always had done. Dad's last day at the windmill was Christmas Day. He acted like it was a normal Christmas, but a few days later he was unable to hide the constant pain. They managed to find a hospital bed for him on January 7th after I complained to the incompetent GP that nothing was being done by the sorely lacking so-called health service that had already let him down. Jacqui and Gerry drove him to the Western General the same day. Dad died on the 2nd of February 1995. He was only fifty-eight. Somehow I had dreaded this day since I was a little boy, and my world ended right there.

See the tree, how big it's grown.

Exactly four weeks later Caitlin was born, and what should have been the happiest day of my life was almost impossible to bear without Dad to share the joy of such an occasion with us all. I was supposed to be supporting Arlene through all this and she did well to get through it with me useless and in tatters. I tried so hard to deal with an emotional cocktail, a cruel blow and then a wonderful gift, but I felt blinded and cornered, cheated and lost. Nothing as trivial as bricks and mortar mattered to me anymore, whether in the shape of a windmill or not. I cannot describe how I felt from then on, but somehow I had to get it together because Caitlin was here now, and things were expected of me that I knew I had to try and live up to. Without knowing it, I think she helped us all through a very tough time; a painful time that for me at least, has never really gone away.

The Old Windmill in 1999 (photograph courtesy Paul Graham)..

13 Poltergeists

Somewhere in the blur of the past six months I had given up the job at the American school in Aberdeen, rejected the anti-depressants that had made me worse and immediately taken up a post as Head of Music at Langholm Academy in the Dumfries and Galloway region of all places. This meant three hours at the wheel every day and it had seemed like a good idea at the time, thinking at least that I was able to come home every night. The job itself was fine, better than fine, but the travelling soon became tedious. Out of the frying pan, into the fire. It took a further year of commuting before I finally managed to plot my way 'home to Edinburgh' after a total of nearly five years in the wilderness.

Since moving into The Old Windmill, we'd had more visitors than the Scott Monument, or so it sometimes seemed. The novelty of living in a windmill hadn't yet worn off (and still hasn't) and every weekend friends from the town would take a nice trip out into the country to see this strange house, often staying over, and for months it felt like we were running a bed and breakfast. By this time, of course, the main attraction was Caitlin, and a steady flow of friends and family continued to head for the hills. Those who came to see the house for the first time were obviously interested in its history, and it was never long before the question arose. Did we have a ghost?

I think so.

The first night that I spent on my own in the windmill was in the early days when the house was half refurbished but habitable, and Arlene for some reason had to stay over in Glasgow. It was only at the last minute that it dawned on me that this particular night would be my first alone here, and I braced myself for an evening when, if the poltergeists didn't get me, my imagination surely would.

I have spent many grim nights alone in desolate, scary places. The old caravan I once lived in had been particularly horrific in the dead of night, and I had often found myself sleeping over in friends' so-called 'haunted' cottages in the middle of nowhere when I was a student, so this should have been nothing to worry about. I loved the windmill and I was not frightened of it, nor was I scared to be alone. I didn't even really have a fear of the dark, within reason. However, on this particular night, I was heading upstairs to bed after doing the usual rounds of switching off the lights and the television, when I was suddenly gripped by fear. Having just left the living room, closing and locking the giant bevelled door behind me with a big iron key (force of habit even on automatic pilot), I turned to pick up a book I'd forgotten, but I felt unable to go back in there. Someone, or something, was in the house.

The whole place was silent and I was completely alone, apart from the two cats who had been lying motionless at the top of the stairs for the past twelve hours, an average day in the life for them. Here, there's never a sound from outside traffic or aircraft flying overhead and there's never a glow of light because there are no street-lamps. It can be seriously scary if you think about it too much. I thought if I could make it back into the living-room, I could switch on the TV and create an ambience of buzzing activity, as if other people are really there with you, such is the power of television – but something still prevented me from going back. It was not a physical thing, but nevertheless a force which terrified me and urged me to go upstairs to the bedroom. I was too petrified even to walk ten yards along the hall to the kitchen to lock up there, so I went straight upstairs. My intention was to get into bed quickly, where I would think happy thoughts and hide under the duvet until I fell asleep, fully clothed if necessary.

With every step I was convinced that I was not alone, and I began to hear the faint sound of voices which seemed to come from outside. Although I couldn't actually hear what they were saying, I was now convinced that people were outside in the garden, talking, babbling, whispering. Men's voices. I was listening extremely hard because they were so faint, and I wasn't even sure that they were real. By this time my behaviour was probably irrational. I got down on my hands and knees and crawled upstairs underneath the tall hall windows so that I would not be seen by whoever was out there in the garden.

I thought about phoning the police, or anyone, but I went through the scenario in my head first, and in doing so I was murdered from behind the minute I started dialling, by someone I couldn't see, so I wrote off that possibility right away. There were no curtains in the house at this stage and I knew that anyone could be watching from the woods. I was sure that someone was out there, or worse, as I had first feared, in the house. I had one more lamp to extinguish in the upstairs hall and it crossed my mind to switch it off and make a dash for the bedroom where I would pull the cord to switch on the bedroom light. But I couldn't face even a few seconds of darkness, so I decided to creep into the bedroom first, put on the light there, and then go back to put off the light in the hall.

I stepped over the cats who were still lying there as if tranquillized, but not quite asleep. Four eyes (maybe more) followed my every movement. They knew that at bedtime there were two possibilities for them. One was that they would be ushered out of the house at forty miles an hour via the cat flap in the kitchen, accompanied by much shooshing and handclapping, or they would be left alone to lie all night in paradise with the hot pipes running under the floorboards beneath them. It was a lottery for them and there was no rhyme or reason to our decision, which was the fun part, and the cats, being intelligent creatures, also knew that the best thing for them to do was to keep a low profile. They had learned a long time ago that if they were to suggest food they'd automatically be out. If they wanted their tummies rubbed, the same. They knew the only chance they had was to lie still and do and say nothing, and with any luck neither Arlene nor I would have the heart to turf them out in the cold. As it was, on this occasion I was more than happy to have them quite close to me for company. There was no way I was brave enough to go downstairs again anyway. Tonight, the cats were staying in.

I made it to the bedroom, my heart in my mouth, and I pulled the cord. The dull forty watt bulb, unfurnished as yet by any lampshade, didn't exactly create a warm atmosphere and by this time my knuckles were white with fear, so I went one stage further, still hearing things, seeing things and imagining things as I sneaked towards the table lamp by the bed. A floorboard creaked at the other side of the room and I stopped breathing and willed myself invisible. But this very act made the eerie silence even more pronounced and I panicked only to hear myself gulp involuntarily as I stood in anticipation of being grabbed by a hairy hand. I waited for probably about a minute, literally scared stiff with my eyes bulging out of my head, looking around like a periscope, but it didn't happen, and my joints loosened up a little as I slowly reached for the lamp.

Click.

The lighting situation was better now, but I could still feel a presence and a coldness in the house and my heart was thumping much faster than usual. I decided to retrace my steps to the hall as planned, to switch off the trail of lights behind me, mainly because I actually felt more frightened being able to *see* along the corridor than I would if the light was off and the door was closed. The cats were lazily watching my return journey, still pretending to be asleep. I knew what they were thinking.

'Please don't kick me out.'

Not likely. I had no intention of removing my only company on fright night as I tiptoed slowly back along the hall to turn off the light. I had barely stepped out of the bedroom when both cats suddenly leaped three feet into the air, tails swinging, claws out, fur vertical, screeching manically. They did a forty-five degree turn in mid-jump and landed a split second later, frozen into position, both mouths wide open, hissing like snakes and growling like rabid dogs, with all four eyes fixed upward on the same thing.

Whatever that thing was, I could not see it.

The cats' eyes never moved off target and they never stopped hissing and growling. I nearly swallowed my tongue as I stood back and waited for something to materialise. The cats stared for ages at whatever they could see and I could not. They were riveted to the spot with fear, but they were ready to fight their hostile apparition. By this time, to add to my trauma, I was now equally scared of these monster cats that I did not recognise as my pets, and I sheepishly stepped back into the bedroom, quivering, as I watched this creep show. I was still hearing incoherent voices and the presence I felt was all but visible. After nearly a minute of this, Smith and Jones very gradually calmed down and slowly began to look away. In front of my eyes they changed from Satan's wildcats back into my domestic ginger toms. They had become pussy cats once more.

Within a few seconds, everything was back to normal. My fear subsided in an instant as if someone had turned off the fear switch. The idea of sensing a presence in the house suddenly seemed absolutely absurd, and the mystery voices had disappeared altogether. Quite rationally I explained to myself that the voices could have belonged to late dog-walkers, for they are a dedicated bunch and when Fido says jump, the owners generally ask how high.

That night, Smith and Jones never did venture outside, and the three of us awoke the next morning to the sensational sound of the dawn chorus, under the covers and all huddled together in the same bed. In the light of a beautiful spring day everything was alright. Scaredy cats.

On several other occasions reports of strange goings-on could be heard over the breakfast table, when our townie guests would make light of the previous night's experience, each in turn putting it down to their unfamiliarity with such dead quiet, and usually admitting that their imaginations had played tricks on them in the black silence of the countryside at night. There was a different atmosphere in the morning anyway, and it always seemed to be bright and sunny and blue-skied, in complete contrast to the night before. I could tell that most of our guests felt a bit foolish and over-dramatic if they even hinted at a spooky tale, in the same way that a country bumpkin like me might whinge about not being able to get to sleep for the noise of the traffic when staying over in the city centre after a night on the town.

One guest was adamant, though, that what he heard was not a figment of his imagination, not whisky or vodka-related and not a dream. Arlene's brother, Stuart, and his girlfriend, Angela, had come for a visit one weekend after most of the work had been done. We spent a night at the Edinburgh Festival, after which we ended up at some trendy basement party, rubbing shoulders with Edinburgh's most pretentious press representatives, disc-jockeys and schmoozers. It was a crashing bore, darling. Clambering over the city's finest journalists and broadcasters, who lay drunk and slumped on the stairs of the exit, some attached to weeping tarts, we headed for home at about two in the morning and went to bed after a Southern Comfort nightcap.

When we all got up at noon the next day, Stuart told a convincing tale of how he had been awakened in the middle of the night by a woman screaming in the woods. Stuart and Angela were sleeping in what is now Caitlin's bedroom at the back of the house which looks deep into Windmill Wood, a dense forest of tall trees like something out of The Three Little Pigs. Among the trees, overgrown with brambles and briars, are bricked-up entrances to the labyrinth of disused mineshafts belonging to 'The Windmill Pit' which had been worked for nearly two hundred years, invisible even to the children of the late twentieth century who play hide and seek in the thicket. At night, nothing *inside* the house can be heard from this bedroom, including the television in the living-room, the radio in the dining-room or any activity throughout the house. Instead, there's a peculiar isolation that lets you hear all the *outdoor* noises quite clearly, like when you're on a camping holiday, and the most common sound is of the wind rustling the leaves on the trees. On a really windy night, the racket can be louder than the traffic in Slateford Road. I knew this because we had spent our first months virtually living in this very bedroom, although I had never heard any *mysterious* noises then myself. Stuart had dropped off to sleep as soon as his head hit the pillow and fell into his usual deep slumber which normally lasts until the middle of the next afternoon, every day for one month out of two. The other non-sleeping month is spent offshore on an oil-rig in the North Sea. If he could survive all that without a teddy bear, then this should have been a walk in the park.

But Stuart had been so convinced of this being an authentic screaming lady during the night that he propped himself up on one arm, shook his head as if to sober himself up and listened more intently. Sometimes sounds can be deceptive and Stuart tried to match the screams with images of big ugly birds that can't fly and live in the woods. He'd seen them earlier in the day waddling around on the other side of the fence. The screams were too human. He consciously cleared his mind and tried to make the sounds fit mental pictures of other animals. Fighting cats can sound remarkably like babies crying. Perhaps 'screaming woman' was in their repertoire. Again, it didn't fit. By this time Stuart was wide awake and sober as a judge, and he could be more creative with his detective work. He tried to envisage mischievous children out late and up to no good but this was far from feasible, and it even crossed his mind that I might have been trying to play tricks on the guests by wiring up speakers for a prank using BBC sound effects tapes. As if I had the time to play such fun games, but I'll bear it in mind for the future. The acoustics were so convincing that the only explanation that Stuart could come up with was that he *had*, in fact, heard a woman screaming in the woods. Throughout all of this, Angela had heard nothing at all, as she dreamed happy dreams in her rollers and blindfold, occasionally snorting like a rhinoceros, but Stuart was having none of her pooh-poohing as he continued to proclaim his unshakable belief in the screaming woman for the rest of the day.

One dismal day in November, I was wandering aimlessly around the garden when an old man, about seventy-five, dressed in head to foot dark green rainwear, accompanied by an equally old dog (in dog years) appeared at the gate on his way for his daily eight-hour stroll around Midlothian. With no real schedule to stick to, he stopped for a blether about times gone by. The dog knew the sketch and put itself down on the tarmac with a thud, panting and overfed, slowly being killed by the old man's kindness.

I have always relished listening to old people's stories of their youth. By listening to my Granny's generation of the family talking about their memories, I have retained clear visions of how I think life was when *they* were young, and tales of *their* own grandparents. If you think about it, it's as close as we get to time travel, and it should be tolerated, cherished and passed on. So I was happy to spend the time of day with this old man, and sure enough, I learned a few things about the history of the area, and perhaps even an explanation of Stuart's frightening experience.

Edgehead now shows little or no trace of ever having been a bustling mining village, but if you look closely in the right places you can still stumble across some of the evidence. To get to the colliery site from the village or the road, the only main route was past the windmill, which stood derelict and unwanted on a small mound in the woods, and heavy, oily lorries at one time must have carried tons of coal past it down that road. The Old Windmill was a common landmark for the miners who passed by it every day on their way to work. It overlooked a large flat patch of grass where the men would play bowls on summer evenings after a visit to one of the pubs in the village, and the old bowling green now forms part of our garden, providing enough space for a good game of five-a side-football or some lumpy lawn tennis.

Of course, the miners, too, had their stories to tell. Tales of hard graft, community spirit and every so often, tragedy. My own great grandfather, John Hall, was killed in Smeaton Pit in 1918. Mining disasters have been commonplace since man first mined, and Edgehead had its share of these during the time the pits were open. The old man at the gate spoke of the drift mines and where they were situated, and he had an impressively clear memory of when they were closed for good. All the pit machinery was finally dismantled in 1958. He described to me a day in the life of an Edgehead coalminer at the beginning of the century, his working conditions, his social life and his home life with his family in one of the little cottages, now probably occupied by an accountant or a footballer or an actor, and passed on a story about a man who had worked in The Windmill Pit with his own father.

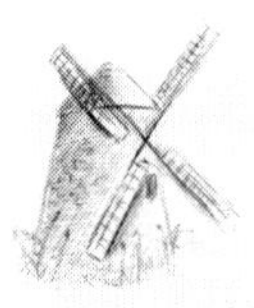

Robert Thomson was a man in his late forties and had worked in coalmining all his life. He had married a beautiful lass (old-man-speak) called Jessie Gray from Humbie when they were both teenagers and they had three sons who had followed their father down the pit. Thomson was a big happy man who lived for his family and always put aside the temptations of the many pubs in the village in favour of saving a few coppers to take the family somewhere nice occasionally. He would save up for day trips to the seaside in a steam train, he would take the boys to the fish markets in Fife, and on a Sunday after kirk the family would go off on a picnic, sometimes even going into an hotel for afternoon tea. As the boys grew older, their father would take them fishing, or bowling up at the windmill and to see the Hearts 'get beat' in Edinburgh. When the time came for them to leave the school, jobs were set up for them at the pit, the management knowing that if they were of the same calibre as their father, they would be honest, hard-working and reliable. Jessie never liked the thought of her sons, or her husband, working in the pit but it had become a way of life and this was the only community they had ever known, so she tried very hard not to think of the dangers.

One night in the middle of winter, a breathless young boy came knocking on the doors of the cottages in the village. There had been an accident at the pit face when some of the timber supports had given way, and the mine shafts had caved in, trapping some of the men. Nobody knew how many men were involved, nor who they were, but Jessie left the cottage in a hurry, grabbing her coat on the way out. Everyone in the village was already running up the hill towards the pit. All the doors in the street had been flung open and left ajar as people, mainly the women of the village, made their way to the scene of the disaster. At the pit head there was utter chaos. Rescue services were inadequate by modern standards and men struggled with monstrous broken beams, pitch darkness and lack of information. Before the coal dust had settled, ten men emerged from the shaft, coughing and spluttering, injured and confused, while the rescuers headed for where they had just come from and started digging for dear life, using their fingernails if they didn't have tools. Meanwhile, the women who had arrived running from the main street were desperate to help, and above all they were anxious to find out if their men were safe. Down below, the exact location of the disaster was pinpointed and another team of heroes went into action, eventually bringing to the surface a dozen more distressed miners. There were cries of relief, and then wails of despair as somebody's son or somebody's brother wasn't among those brought out. Ropes and gas lights and stretchers and blankets were brought to the scene and people prayed and held their breath until another team of rescuers came back. Families were stricken by sheer worry for the men who hadn't come up yet. This went on three or four times until every man who could make it was safely at the top. There were regular head counts while the rescuers continued to dig and scrape. They defied the swallowing black seam as they went down again and again in search of more survivors, or at least a better assessment of the situation. The work never stopped all night, with little accurate information coming through and no communication with those above ground. Then, at last, silence. Everyone was hushed quiet as the men who had taken charge listened for voices underground – a cry for help, a groan, anything; but there was nothing. Then they would dig again in the collapsed shaft just in case they found someone, but in the confusion they didn't even really know who they were searching for, if anyone at all.

After nearly ten hours, as the new dawn broke, the rescue teams were satisfied that they had done everything possible, working only by the dim light of their pit helmets, and the call was made for the remaining rescuers to return. As the men walked, completely exhausted, from the shaft unaccompanied by any new casualties, they were embraced by their own wives, colleagues and villagers who had stayed to help and comfort people.

Then there came the semi-silence of the aftermath of a disaster, and among the anguished and dirty faces, one woman had nobody to hug. Jessie Gray panicked in the grim realisation that no-one else was expected to be found alive, and her husband and her three boys were the only ones unaccounted for. Everyone seemed to realise this at the same time, and a stunned cold quietness fell on the devastated community as people stood helplessly still in the early pink sun. The silence was broken seconds later as Jessie screamed uncontrollably, weeping for her family, the only four men to be lost in this mining tragedy. Their bodies were never recovered. The old man at the gate told me that Jessie Gray is said to have lost her mind, returning to the pit head every single day until she was very old and frail, telling neighbours that she was off to meet Robert Thomson and her boys after their shift, and each day when they failed to walk from the mine shaft, she fell to her knees, screaming and crying inconsolably.

Perhaps this story explains Stuart's screaming woman in the middle of the night. More spookily, though, there have been numerous reports of sightings of a cluster of dim lights approaching in the distance as you walk towards the woods on the old pit road past The Old Windmill, in the dark. The lights approach slowly but remain absolutely vague and indistinguishable until the very last moment, when those who have seen them swear that beneath the lights, the figures of four miners appear, as if walking home from the pit towards the village. The distant lights turn out to be the lamps on their pit helmets, and for a moment the whites of their eyes can be seen glowing against their coal-blackened faces, as real as you like. Then in a split second and a turn in the road, the four miners vanish into thin air.

The old man went off at a tangent and it was difficult for me to keep up with the people he was talking about and in which period in history we were, but his memories were vivid and I got the feeling that it wouldn't take much effort on his part to conjure up his own past and to see all the familiar dirty faces walking this road, when it was a hive of industry, as if it was yesterday. I have often noticed that it is not uncommon for old people to relate their stories with tears in their eyes, clearly touched by their own vast memories of people and things from another time, another place, who are gone forever. I listened attentively until the old man decided to move on. He wiped his tears away and waved his walking stick high in the air as he picked up speed and strolled off at a good pace. The old dog prised its rear end off the ground slowly and painfully, and limped away behind its master, following him deep into the woods, on the path of the old pit road.

Edgehead: the mining village.

The site of the Windmill Pit as it is today.

Everyone loves a good ghost story, and I now know many associated with The Old Windmill and Edgehead pit which are believed to be true, but I can only really vouch for my own personal experiences in the house, which include the cats' tale, the occasional uninvited voice or presence, and one final incident which freaked me out.

I was alone in bed in the middle of the night, having done the rounds, switching off lights, doing the security checks and turning the big iron key to lock the living-room door, the usual routine. Arlene was working away for a couple of days and I had enjoyed a few hours of deep sleep, when I was suddenly aware of someone tapping me hard on the shoulder, just once, as if they were trying to wake me up. Not only did this interrupt my pleasant dream, but the dull pain remained long enough to cause me to have to rub my shoulder, such was the impact. At first I thought Arlene had come home unexpectedly and was trying to move me to my own side of the bed, but she hadn't. I was lying in the middle of a giant bed with no objects around me to bang into or to fall on me. As I lifted myself up on my elbows in the darkness to see who was there, someone closed the bedroom door behind them on the way out. I thought I was being burgled so I buried my head under the covers and froze with fear. Somehow, I must have fallen asleep almost instantly, sparing myself a night of terror, and in the morning I tentatively crawled from under the blankets and went back over the events in my head, still convinced that it had been real. When I went downstairs all the windows and doors were triple-locked as usual, and obviously there had not been an intruder. The cats, on this occasion, were locked safely in the kitchen, where their cat flap to the outside world lets them come and go without giving them access to the rest of the house. But on my way back to the bedroom, I registered the fact that the living-room door had been unlocked (not by me), and further upstairs the hatch to the Secret Room above was also, rather mysteriously, ajar. More convincingly, I had a small bruise on my shoulder for a week, where I had been prodded. Sweet dreams.

Until recently, the road sign that proclaimed your arrival in Edgehead, just after the thirty mile an hour speed limit sign, not only read 'Edgehead', but in brackets underneath also gave the name 'Chesterhill'. So, we had a choice. We could say our village was called either Edgehead, or more historically, Chesterhill, and neither was wrong. However, the letter 's' and the last 'l' had been removed from 'Chesterhill' on the road sign. It's one of those juvenile pranks that keeps boredom at bay on long light nights during the school holidays. On this occasion, I expect the Neighbourhood Watch Committee caught the culprit in the act of removing the letters and had him shot at dawn. For the same reason, motorists can be forgiven for thinking that the nearby town of Pathhead exists only on the map, because all signs in the area lead you to a place called 'Fat head'. Hilarious.

On the same subject: my sister, who is a talented linguist, an honours graduate and a promoted teacher of modern languages, had driven past the Edgehead-come-Chesterhill sign many times on her way to the windmill. One day, the old sign was suddenly replaced with a new one on which all eleven letters were present; it also incorporated the thirty mile an hour speed limit (bigger and bolder) and offered a friendly 'welcome' on behalf of the Village People. On seeing the *new* sign for the first time, Jacqui bravely admitted this to us:

She had seen the old sign every time she'd come into the village, but had gone through life thinking that 'Che terhil' was a village in France, twinned with Edgehead, as Jarnac was with Dalkeith! The realisation that Edgehead had another name, 'Chesterhill', took her completely by surprise. Quite rightly, Jacqui has heard no end of slagging about this, especially from Arlene; and when *this* happens all Jacqui can do to retaliate is bring up the quail's eggs story. You know, the one when:

Arlene and I were invited, as newcomers to the village, to a wonderful reception by the Rankins at their Scandinavian cottage at the Fuffets, about a mile or two from 'Che terhil'. Ron Rankin had been most welcoming and helpful to us in the early days, and he'd upgraded our alarm system. He and his wife had arranged a wee welcoming party for us, inviting some of our new neighbours so we could meet them for the first time, and Mrs Rankin laid on a generous buffet, from which she offered Arlene some quails' eggs in beautiful mayonnaise, on a pair of delicate tweezers.

Arlene's momentous reply to this was, 'No thank you. I don't eat fish.'

Mrs Rankin politely explained that a quail was a bird, and Arlene made a futile attempt to cover her tracks by saying, 'Oh, *QUAIL!* I thought you said *WHALE*'.

At this point, the vision of a whale's egg being offered up on tweezers as a delicacy flashed before me. I choked on my pineapple and mushroom vol-au-vent and died quietly of embarrassment in the corner, because by this time a JCB could not have dug Arlene out of that one.

14 The Hit Squad

I wandered around the garden kicking stones and pine cones off the grass. We certainly had our work cut out for us if this garden was to live up to the expectations of the 'Scotland in Bloom' wing of the village committee. The competition was an annual event and Edgehead was a serious contender. Two summers ago we had broken Clara Chalmers' age-old tradition of inviting the judges for 'tea on the lawn' at The Old Windmill while our work on the house took priority, but we felt that this year we might at least be able to make a small horticultural contribution. Last year, by chance, I had looked out over the park on judgement day from the top room of the tower, only to see the judges being awkwardly ushered away from our house by the hosts from the village committee. It seemed that if they got too close, we would blow their chances completely. As it turned out, Edgehead still won a very respectable second place that year, with all credit going to those from the village who seemed to work tirelessly from dawn till dusk, as they do each year, keeping the main 'drag' in particular looking like something from Gardeners' World.

This year would be different. Although tea on the lawn for the judges was never likely to be a possibility as long as we were in residence, we did have slightly more time on our hands (even with a child) to make a little more effort this year. Actually, laziness prevailed in the end, as an ingenious plan came to us. 'The Beechgrove Garden' had been a long-running BBC television series on gardening, and a popular weekly feature on the programme involved the 'Hit Squad' visiting people's disaster areas and turning them into attractive gardens. The very thing. I watched The Beechgrove Garden with interest for a few weeks to see if I could establish any rules or special criteria that we might need in order to qualify for Hit Squad treatment. It looked like there were two main requirements. The garden should be in a hell of a state, and there should be an interesting angle such as:

> *The house was once a windmill, built in the eighteenth century; its original architect and owner worked for Auntie Beeb (Elizabeth Adair was the first female voice of the BBC in Scotland) as did its new owners (Arlene continued to work in BBC Continuity and I had just signed a contract for two and a half years' worth of weekly music programmes).*

These tenuous links were dream material for television producers, and The Old Windmill was a dream location, and after a short reconnaissance visit to the site, the producer agreed that the Hit Squad would come to sort us out in June of that year.

At 7 o'clock one morning in June a ridiculous garden shed on wheels, the size of a Luton van, drew up outside our gate. This was the Hit Squad's promotional vehicle which drew a lot of attention at garden fetes, publicity stunts and the like. As it turned out, most of the team were too embarrassed to be seen in such a silly vehicle and they preferred to make their own way to Edgehead, so the road outside the house was soon blocked with a convoy of Golfs and Cavaliers driven by cameramen, sound engineers, a producer, his assistant and the Hit Squad themselves, Jim and Walter.

The early-bird dog-walkers soon sent the jungle drums beating that the Circus had come to town, and for the rest of the day there was a steady stream of pets being walked, as their owners came to catch a glimpse of our messy garden slowly changing into something that would hopefully not embarrass them at this year's Scotland in Bloom contest. The two presenters, television personalities Jim and Walter, are credited with doing most of the work, and indeed they did. They planned out what they were going to do, they gave us advice about which bits should be tackled and they made up funny and informative things to say on camera; but what the viewers don't see is that everyone else has to muck in too in between recording shots of the ever-changing garden, because deadlines have to be met. For example, the cameraman spent more time strimming than filming, the director spent more time hoeing than directing and the sound engineer swapped his microphone for a spade for most of the day. Jim and Walter looked bewildered at first at the prospect of tackling what they saw before them, but then again you could see that in a sense they had the keys to the sweet shop, because there was so much scope for enthusiastic gardeners like them to make the most of rescuing our existing plants and trees as well as to add new ones to the setting.

The first thing to be sacked was our giant border which stretched from underneath the kitchen window at the front of the house into the middle of the drive. It had originally been a beautiful rose garden but time had taken its toll and most of the roses had long since disappeared. Those that remained looked sad, shabby and lonely. A fragile bird table made of logs stood in the middle of the rosebed like an old scarecrow, and would have collapsed into twenty bits if a sparrow had ever come in to land on it. The sheer size of this no man's land also meant a detour of fifty yards if you wanted to walk to the garage, and in its present state it just seemed like a waste of space. Walter took the border back to within a foot of the house and marked the boundary with 'barley sugar' edging, overlooked by a new trellis which he screwed to the wall between two windows. He then planted a variety of colourful and unpronounceable items, including a gorgeous blue thing that climbed its way up the trellis. (You must excuse the absence of any proper horticultural terminology.) The surplus earth from the old rose garden was moved in a convoy of wheelbarrows to the other end of the operation and used partly for planting three new trees in front of the fence to cover up a particularly exposed area, and to offer us some privacy here. Walter's new border was the first step to a striking change and it already gave more of a welcome to the front door, which was enhanced even further when he trundled from the 'shed van' three large terracotta pots that Bill and Ben could easily have lived in. A spot of flower arranging, and Walter had made us the envy of the village. More gravel chips arrived courtesy of Knight Bros, where my Mum had got us more discount, and when they had been spread out there was room for at least another two cars in the drive, although that was not a priority.

Would you be seen dead in this vehicle?

The Hit Squad standby for action.

Meanwhile, Jim was hugging trees. Jim loved trees and he was a tree surgeon first and foremost from what I could gather. He was certificated in using a chainsaw the size of a Honda 125 and soon the Edgehead Chainsaw Massacre went on noisily around us, at altitude. Jim paid due attention to dozens of trees and was happy that he had been able to service them, at least as a temporary measure, before they fell into a state of neglect, but there were far too many for him to deal with in the one day when there was so much other work to be done for the cameras.

While on the job, Jim fell in love with a very old lady called *Prunus*; he cut off some of her limbs and enthused about the tiger-skin markings on her trunk. Sounds like a beauty; whatever turns you on, Jim. But this old tree *was* a beauty and one of the highlights of the garden. Unfortunately her trunk was hidden under her own canopy of leaves and stuck behind some other overgrown bush, a conifer that was supposed to be three feet high but had changed its character and become a climber as it fought for light until it was literally ten feet tall. I, for one, didn't even know *Prunus* was there until Jim pointed her out, but within half an hour, she was the focus of that part of the garden, brought to life again for all to see.

One step down from *Prunus*, in a newly discovered bed, Jim planted what he described as the best variegated bamboo, better known as *Arundanaria viridistrayata*, obviously. A crowd from the village gathered in the drive as Jim did a link to the camera while planting this – enthusing to the viewers, in particular, about its golden variegated foliage. He then walked further up the bed, still gesticulating to the camera as he arrived at his newly interred North American maple, also known as 'the flamingo' and whose colours went from cream to pink to yellow to green. Jim said this would grow to ten feet, but he instructed us, with an eternal reminder through the magic of television and video, that we must keep the flamingo trimmed to just six feet, only pruning it when dormant, otherwise it would *die*. It was a sobering thought that this poor helpless creature's life was now in my hands, and I have been paranoid about its welfare ever since, Jim's immortal words ringing in my ears every time I go out the front door.

Jim's idea was well thought out. The plan was to enjoy a three tier aspect. First, the *Arundanaria viridistrayata*, remember, with its golden variegated foliage and three feet in height. From there the eye naturally follows on to the North American maple, or 'the flamingo' (pruned to six feet but only when dormant or it will *die* and it'll be all my fault) and finally the big one, the *Prunus*, at forty feet. 'A graduation in elevation', as Jim put it, one complementing the other.

As I jumped out of the way of television cameras to avoid being captured forever dressed as a flowerpot man, Walter by this time was hacking an enormous quince from the side of the tower using a pickaxe and another tool that looked like shears on stilts. I couldn't help thinking how naked the building would look if Quincey was removed, because this was the first thing you saw as you approached the windmill. So with some apprehension, I watched Walter strip the wall of its sprawling wall shrub. There was a pleasant surprise, however, because underneath the messy quince there appeared a striking *Hedera helix*, which I recognised as 'ivy', perfectly formed and clinging to the stone in greens and yellows, and known as 'Goldheart' according to Walter. That wasn't so bad after all. In fact the improvement was dazzling, like a bearded hippy with new smart haircut and a shave.

The work went on literally from first thing in the morning till last thing at night. Arlene and I made a token contribution transporting wheelbarrow-loads of garden debris into the woods; one or two children from the village lent a hand on the off-chance that their pals would see them on the telly, and Mum made gallons of soup for the troops and kept everyone in tea and sandwiches all day. Caitlin, aged three months, made her television debut in one short scene (this saved us sending photos of the child to friends and relatives up and down the country) and Arlene did the talky bit at the beginning of the programme (where you have to look desperate and hopeless) and the bit at the end (where you can't believe how marvellous it all turned out). I was asked to take part in these sequences but I declined on the basis that I have too many rock'n'roll friends to be seen on a gardening programme. In the end they managed to sneak into the final edit an unflattering shot of me which lasted only a fraction of a second. The irony is in the number of my rock'n'roll friends who, it seems, watch gardening programmes, as nearly every one of them phoned me up to say they had spotted me dressed in silly red socks and baggy shorts on The Beechgrove Garden. It was certainly a day with a difference and the celluloid evidence remains in the archive for posterity. I almost didn't recognise my own front garden when the Hit Squad had finished. My only regret was that I had to leave about tea-time to drive to Inverness for a gig I was playing with The Colour Wheel to promote our new album, although I must admit I would have preferred to stay in the garden to the end and enjoy the crack. How rock'n'roll is that?

The pressure was on us to maintain the garden in at least the condition in which the Hit Squad had left it. They had spent a good bit of money on exotic plants, huge pots, tons of gravel for the drive and more trees, including a holly tree that Jim planted beside the pond in Dad's memory. We always feared Jim or Walter might turn up out of the blue one day to find half of the plants and trees in a state of neglect, but luckily this hasn't happened so far. Of course they'd be made most welcome, and immediately handed a rake or a trowel. Fortunately, we did manage to keep on top of the garden that summer, without going overboard, and Edgehead won the Scotland in Bloom competition for the next *two* consecutive years (at least!).

The Hit Squad create a warm welcome.

15 Landscaping

With the bulk of the work now done, I was beginning to afford myself a bit more breathing space, and I would find myself wandering around the garden at my leisure, observing the differences between the converted windmill of 1960, well before I was born, and its newly refurbished state thirty-five years later. Apart from the success story of the house, I felt fortunate in that we had inherited a mature, well-planned garden with such a variety of shapes and colours. I could only describe it in terms of shapes and colours and sizes as I knew none of the lengthy Latin multi-syllabic scientific names. What's more, I was a bit ashamed of myself when I recalled that when I was a student I had worked *five summers* as a gardener with Midlothian District Council, where occasional opportunities to learn about plants and flowers certainly arose, but I had other things on my mind then and I was just as happy to lean on a hoe for hours on end outside Easthouses Community Centre with the other members of the chain gang instead. I rue the day I didn't listen to the old man with the green fingers in the crystal palace that was the massive greenhouse in Newtongrange Park, because he was willing to teach me things I would appreciate now.

As a result, to me, the *Tilia platyphyllos* or large-leaved lime, is 'that hairy tree with the yellow flowers' and a *Laburnum anagyroides* is 'the thing that looks like a firework'. Having said that, I do remember the beautiful copper beech, or *Fagus sylvatica* to you, from primary school and I can tell a dogwood tree by its bark. Our own big copper beech looks wonderful in the morning when you get a perfect view from the bath, its fantastic colours metamorphosing before your very eyes. When the tree was young, in the days before whirlies were invented for hanging out washing, one of its previous owners hung a washing line around its trunk, tying one end of the white plastic-coated wire around the tree and fixing the other end to a hook on the garage wall so they could dry their clothes in the sun. By the time we came to live here, the tree had grown so much that the trunk had completely swallowed the clothes line which was wrapped tightly around it. The wire had very gradually sunk into the bark, at first causing a deep imprint in the tree, but later becoming impossible to separate from it. We were advised to cut the line free as the constant tension on the trunk would damage the tree in some way, but there was no way of retrieving the remains of the washing line, which you can still see sticking out of the bark today, impossible to budge and slowly disappearing further into the tree as it grows bigger over the years.

Other trees in the garden may have been planted too close to the wire perimeter fence at the back, because some thirty years on, some of the larger trees are now straddled by the fence, having grown up through the middle of the lengths of barbed wire which have now sunk into their trunks. The Hit Squad had to surgically remove one of these trees to make room for another which stood too close to it (a bit like dentistry, I imagine) and now a sizeable chunk of wood hangs impressively from the top wire of the back fence as a kind of unusual feature, I like to think. Let's see you grow one of these, intentionally.

We continued to keep in touch with Elizabeth Adair, who still seemed to show an interest in The Old Windmill after all those years. She had been away from the place much longer than Arlene or I had been *alive*, so I was impressed with her memory of it all. When the BBC's Hit Squad came to give us the makeover, she was invited to participate in the programme, but was unable to do so due to failing health, and I know that the producers of the programme were disappointed about losing out on this particular angle.

I kept thinking that Elizabeth Adair should see her garden in full bloom, mature and vibrant, even though many other interesting things had occupied her life since the early sixties and The Old Windmill episode was now probably just a very fond memory from a very long time ago. But I couldn't help thinking it was more than that. She had fallen in love with the building and the location; she had been moved to write a book about her experiences here, and from reading the book I could tell she was obviously passionate about the whole project and everything that surrounded it: conservation, the countryside, nature, birds, plants, trees, animals; I even got the impression that the garden may have been more important to her than the house. It's easy to forget how much she actually had to do. Faced with nearly an acre of jungle on the edge of a wood, Elizabeth Adair persevered with a flair for garden design and landscaping. She had sought good advice on how to tackle her new house's prospective garden, partly from the East of Scotland College of Agriculture (who offered a free service at the time and sent a knowledgeable representative round to help), and consequently everything was carefully chosen for its beauty, contrast and variety.

It had taken a careful concoction of different grass seeds to transform the plateau of rich brown earth, ploughed by the local farmer as a favour, into a smooth lawn on which you could soon play bowls once more just as the miners had done a hundred years before. In other parts of the garden the rough grass mixture comprised such ingredients as Perennial Rye Grass, Chewings Fescue, New Zealand Brown Top Bent and Crested Dog's Tail. To add colour among the larch trees Elizabeth planted groups of bluebells, narcissi, early tulips and scillas. Floribunda originally occupied the giant rose bed we eventually cut back, but there were still traces of clematis and rambler roses climbing the garage walls when we arrived. Twelve trees had been carefully selected by Elizabeth from an identity parade, and delivered from Aberdeenshire. There was a laburnum planted beyond the garage, three silver birch and a thorny bramble hedge. Five evergreens stood guard just inside the stockade part of the fence – the Deodar Cedar from the Himalayas, the Giant Fir, the beautiful imposing Blue Atlas Cedar and two tall slim cypresses; then in contrast, an *Allumii* with unusual grey leaves and the silvery *Argentea*. The copper beech was put in the wooded part to the south and the quince sat for thirty odd years at the foot of the tower until Walter from the Hit Squad tore it out.

Elizabeth Adair's garden in 1960. The young Pyrus stands beside the steps.

The pond under construction.

The garden in full bloom.

The young holly tree overlooks the pond.

One of my favourite trees is the *Pyrus* which was planted at the same time as the others, when it was only eight feet high, on the grassy knoll on which the tower was built. Three steps were cut beneath it into the bank leading to the pond. Nowadays the *Pyrus* is nearly forty feet tall, the same height as the windmill, and its fantastic blush pink blossoms can be seen at close proximity from all three floors, its branches almost reaching into the windows of the house. We were a bit concerned when part of the trunk seemed to turn black last year, so we invited a range of tree surgeons to come and give us a survey and an estimate. The first callers said the tree should come down, which was not what I wanted to hear; the second team gave me a vague analysis that was neither here nor there, and then a third company said the tree could be saved but at great expense, offering us a special deal for them to remove the dead bits from the premises should we decide to axe it. No thanks. Others confirmed that it was fine but it should be treated right away, so in the end we contacted Jim McKirdy from the Hit Squad again, who came at once with his bag and his hat and knocked on the door with a rat-a-tat-tat, and saved the tree with minimum fuss. Now it continues to bloom and grow forever.

The landscaping of The Old Windmill's garden was no mean feat and I certainly didn't envy Elizabeth Adair's part in it all when I read about her constant frustrations. The conversion of the windmill and the construction of the extension were frustrating enough, but the little things clearly ate away at her patience too. The ink-blot pond was a stunning feature in the garden and was made of bricks and lined with water-resistant cement. To this day it has retained the water faultlessly, but while it was being built the gallons trickled away three times before it was made watertight. The Sower, the magical statue who stood overlooking the pond, was stolen by thugs who had to break his ankles in an attempt to prise him away from his concrete base, only for them to find that he wasn't made of lead and was therefore worthless to the thieves, and an early attempt at home security was foiled when a 'no trespassing' sign, written in Elizabeth's own fair hand, was flung to the ground and trampled on.

Many of the trees and shrubs listed in the book were accounted for as I strolled round the garden putting the bits of the five-hundred-piece jigsaw together in my head, but many more must have been added by the Chalmers family, because at the last count the windmill was surrounded by seventy-three trees, from the common to the exotic.

The garden, as it is today, was Elizabeth Adair's vision. It was also her first big project of this kind. I felt she *had* to see how it all turned out. I was sure she would appreciate the garden much more than I was able to in my first few years here, and of course only she would be able to compare its present resplendence to its infancy many years before when she had planned it and planted it, having to leave it all behind before it fully flourished. Sure, it had been adopted and reared by Clara Chalmers for the next thirty years, but now it's all grown up. Sadly, Elizabeth Adair never returned to The Old Windmill.

16 The Secret Room

Back indoors, the fine tuning continued. What to do with the Secret Room was a difficult one. It was a unique, fantastic place with panoramic views, brightly-painted rough stone walls, and a cavernous cone-shaped roof like the canvas red and yellow interiors you used to see when you looked up, if you dared, from the chairoplanes or the merry-go-round at the shows. Access to the room was via the ornate iron spiral staircase and there was a gateway to the large flat roof outside. It was warm and comfortable and cosy, yet also unrefined and industrial, almost as you would imagine the inside of a stone windmill to be. It had all mod cons too: controllable central heating, power points, cupboards and curtains; stripped wooden floors, a patterned rug from the 1930s, a really uncomfortable futon (on which no-one ever managed to get a good night's sleep), two antique chairs, a double bass and that bloody teacher's desk that took forever to put up there. The Secret Room would make an excellent study, or a playroom, or even a guest bedroom. It could serve as an excellent music room, or an artist's studio or a den; even an observatory with a little work. In fact, it could be anything you wanted it to be, and it took years for us to decide what to do with it.

I intended to make it my own recording studio, at least to start with, so I gradually sneaked the components upstairs one piece at a time over a period of six months until I was wired for sound. For nearly two years the Secret Room served its time as a small recording studio while I recorded the preliminary stages of an album I was working on, and when the album was complete, I could proudly give the house a credit in the sleevenotes for its part in the process. When this particular recording project had finished, I decided to relocate the equipment to my office with a sea view from Loretto School, where I was working by this time, away from sticky little fingers which had begun to jam up the mixing desk with cake and cover the keyboards in chocolate. Besides, an increase in the number of weekend visitors to Edgehead often meant a place was needed for them to sleep, without having to wade through ten guitars and a spaghetti junction of cables and leads.

The futon had been a classic bad buy and we gladly gave it away, regretting how much it had cost us in the first place, but if *we* weren't prepared to sleep on it because of its monumental discomfort, I didn't think it was fair to expect our guests to endure the back pain. No wonder the Japanese are tough machines. It was replaced by its extreme opposite: the most comfortable bed in the world: it was a bit of a pantomime getting this lot up the spiral, but well worth it in the end. The replacement was a new bed of black wrought iron with simple wooden posts, much more in keeping with the house than the futon was, and it came with a soft bouncy marshmallow mattress.

Almost overnight, the Secret Room had become more of a bedroom. Arlene impressed us all by painting and re-covering two basket chairs; we had also picked up two classy antique chairs many years before in a moment of impulsive self-indulgence when we had nowhere to put them, in the hope that one day we'd have a suitable room, and they suited this room perfectly. She also made funky curtains for the four windows that complemented the new colours and jazzed up the room even more. A flashy computer arrived so we could work up top in the relative tranquillity while enjoying the inspiring view across the village and beyond. The acoustics turned out to be exceptionally good when the CD player was installed or when musicians turned up for rehearsals. Having no neighbours meant I could crank my music up to eleven. The double bass remained propped up in its corner (relatively speaking), its posture the same as the musician who would play it, and we picked up some other *objets d'art* to make the room even more interesting. I managed to get an old red telephone from the sixties like the one out of Batman, and I brought my old ventriloquist's dummy down from my Mum's loft and sat it at the window to scare any children passing the house. But the ultimate feature in the Secret Room was still to come.

I waited until Arlene's work took her away for two consecutive days, then I commissioned Barry the joiner to sneak in and build a sturdy ledge seven feet off the floor sticking out of the wall by four inches all the way round the eighteen foot circumference of the room, just below where the four foot thick wall meets the slant of the roof. Cork strips, about two inches wide, were glued on the top surface of the ledge and then OO gauge railway tracks made of 'flexitrak' were tacked into place in a perfect circle (not by me, I should perhaps admit). Caitlin and I scanned the local papers under the junk bargains section where I spotted a train set for sale for less than fifty quid. Off we went to a house in Mayfield where a man in his fifties led us up into his loft which turned out to be an Aladdin's cave of toy modes of transport I hadn't seen since I was a toddler. He had them all: the original silver Corgi James Bond DB6 with the revolving number plates and the bullet shield at the back, the Batmobile with its rear jet exhaust pipe still working and its yellow missiles which could be fired from the rear guns, a 1961 Bedford Romany Dormobile with the sliding doors and the plastic roof that you can pull open, Captain Scarlet's SPV: the ultimate driving machine with the side door that opens when you press a button in the roof, and the lesser known white crab-like SMV with the gullwing doors and its precious cargo, the tiny crate of gold with the detachable lid. There was a blue MG 1100 with a plastic man in the driver's seat and a collie dog sticking its nose out the back window, a Hillman Imp with a plastic glass tailgate that opened and closed, and a back seat that folded down when you pushed the car forward under pressure. There was a tiny cream-coloured Radio Rentals van with a plastic aerial, still unchewed, and a back door that rolled up and down, and a gold Jaguar Mk 10, identical to the one my Granny had bought me in nineteen sixty something: I used to get annoyed with her as she drove it around on the carpet *in reverse* all the time when she played 'cars' with me (blind as a bat).

There were cars all over the place in the collector's loft. Some were on display, some in their original boxes and packages, and many lying loose in shoeboxes, but the bulk of the attic space was taken up by an enormous train set. Not being a model train anorak or a

Hornby buff myself, I took little interest in his prize collection and quickly cut to the part where we negotiated which bits I could take away for forty-five pounds. He was keen to give me the lowdown on which wagons should go with which engines and so on, but I couldn't really get into the game. I really only wanted a collection of 'trains' that would look good whizzing around my Secret Room at a hundred miles an hour! With this in mind, I chose an old-fashioned steam engine with 'King Edward I' written on the side, two first-class carriages that seemed to look the part, a restaurant on wheels, a giant red and white Wimpy Hamburger crate, a cylindrical United Dairies milk container and a dull brown guard's wagon that looks exactly like the Edgehead bus shelter on wheels.

After ten minutes in the attic, Caitlin was trying to do a parachute jump through the hatch into the hall below and I was getting stressed. The trains were hastily packed into an old cardboard box along with a transformer and some spare track, and I handed over the money. When we got home I was dying to see the new attraction in its full glory trundling round the room. I plugged in the transformer and connected the wires, helped (no, hindered) by Caitlin, and after a few sparks, false starts and even a couple of serious derailments in the style of the Tay Bridge disaster, King Edward I made his way round and round and round the Secret Room. Caitlin took the controls, and I followed the train closely from down below with my hands cupped and my arms stretched out ready to catch it the next time it flew off the rails. Ever since the opening of the mighty Trans-Secret Room Railway we have seen big smiles on lots of wee faces, including mine.

The Secret Room as found by Dad in 1993.

The Secret Room in 1998, featuring the Circle Line (photograph courtesy Paul Graham).

17 Reflections

Maintaining an old property is like painting the Forth Rail Bridge, or so the cliché goes, and this is without a doubt true. In all honesty, I wouldn't have it any other way but you do have to tolerate the prospect of perpetual maintenance costs as you watch materials deteriorate before your very eyes. You know that one day, probably next year, more window frames will have to be replaced and their prices will be astronomical. The patio has all but sunk into the soil, and while we could quite easily live with it that way, a new one would enhance the garden considerably. The perimeter fence is hanging by a thread in places; only the section that Dad fixed is still secure, and the cost of a new fence is a horrible thought. The kitchen refit was a temporary measure three years ago when all the money ran out and I commissioned Barry the joiner to make some units and bunkers out of scraps of wood left over from other jobs. It's got character, but not nearly as much as the original 1950s kitchen had. It's a shame the original one had to go, but there was no way of fitting essential new pipes, cables and a boiler without demolishing it at the time, or so I tell myself, and now the desire, and hence the cost, of a new kitchen is always looming. Roof repairs will always be a probability, and when they're required we can say goodbye to holidays that year and hello to an (even) older car.

I have always lived in old houses since I started working and I have always been faced with such frustrations, though on a smaller scale, but I take all the knocks on the chin, as I still believe in a sadistic sort of way that maintaining this ole house is all part of the fun. For example, after two or three years of living in Edgehead, we eventually got time to switch on the television for the first time. We had been given a TV by one of Gerry's sisters, and formerly it had belonged to his parents in the early seventies. As I turned the dial on the ancient set (no such luxuries as push buttons or remote controls) I expected to pick up programmes like Screen Test with Michael Rodd, or Clapperboard with Chris Kelly or Ask Aspel. Everything would be in black and white. Dickie Davies might have been on presenting Grandstand or we'd see Anthony Valentine in Colditz with that terrific haunting theme music, or even an episode of Crossroads that went out live.

At last I found a snowy picture of what could well have been The Holiday Programme, but no matter how long I spent trying to tune in the TV set, the picture was virtually unwatchable. We lived with this for as long as we could and eventually we decided to replace the set with a newer model in the hope that we would get a better picture. However, the reception was the same.

I stood at the bottom of the windmill's tower tracing the cable with my eye up to the stalk of the aerial which seemed to be wrapped around the chimney with a bit of chicken wire. This may have been the problem. A week's worth of phone calls to TV aerial companies, who are *never* in, resulted in two estimates for repairs to the aerial. Both companies refused to guarantee a perfect picture at the end of it and each said the chimney was falling down and should be secured first. Money. Expense. Damn. More estimates, this time from roofers, more humming and hawing and more hassle.

Nothing is ever straightforward, and in this case we had to arrange for costly scaffolding to be erected 'to please the roofer who would fix the chimney so the TV aerial man could sort the aerial, and we might get an improved picture, but there were no guarantees'. When the head roofer got to the top of the scaffolding he announced that the chimney stack would have to be replaced altogether. What did I know, and how could I protest? Was this the roofer on the fiddle? I certainly wasn't going up there to check. If the stack was unsafe, the job would have to be done. This meant that the prospect of my busking career taking off through necessity was drawing closer. The roofers diced with death up above the streets and houses, tossing the bricks of the old chimney stack to the ground forty feet below. Meanwhile, inside the freshly-painted, beautifully-refurbished Secret Room, loose plaster and dislodged cement rained down on the newly-opened railway, as I peered at the black sky through a hole in my nice ceiling, unsmiling. As soon as the new chimney stack was in place I wanted these particular roofers off the premises, and I would deal with the interior repairs myself with my own steadier hand, later.

After all that it was time to call in the TV aerial men again, who at last were able to secure a new aerial to the now solid new chimney stack and pass a new cable to the television via a wee white box. Downstairs, Arlene and I waited in anticipation to find out what the people in Eastenders really looked like, and we were curious to put faces to the voices of all the newsreaders. But this wasn't exactly to be a complete success story, even though the man managed to make a very slight improvement to the clarity of BBC1 and BBC2, but only when the wind blows in a certain direction. STV is still non-existent and Channel 4 comes and goes; coming when you don't want to watch it, and going when there's something on that you *must* see. I made enquiries with satellite companies to see if they could suggest a way of picking up any of the four terrestrial channels but the answer was no. I then pursued the cable route, but again there was no opportunity for us here, as cable had not yet come to the village. (Gas hadn't reached Edgehead yet, for heaven's sake, so what on earth was I thinking about?) I was listening to Tomorrow's World on TV one night and I heard the lady say that a new system of digital television 'is on the way soon' and I live in hope that this will enable us to see pictures in Edgehead one day. I feel as excited about the prospect as John Logie Baird must have done all those years ago. For now though, Arlene and I will just have to continue passing the time in the evenings with our usual pursuits of knitting, reading poetry and doing difficult jigsaws of a hundred pieces or more.

Looking back, I realised that in order to improve our BBC reception very slightly, nothing more, we had splashed out on an Eiffel Tower of scaffolding, a new chimney stack, new cables, a new aerial and a new fifty pound white 'box' which does absolutely nothing. I had also lost the will to complain, so I have brought this chapter to a close until the actual advent of digital television.

Out in the garden, things were gradually improving. Mum had given me a voucher for my birthday to have a landscape gardener come in and do some work. Now there's a hint if ever I saw one, not to mention a sign of old age. (Last year I got a lawn mower – next year – a tool box?) Arlene decided on my behalf that my voucher would be used to create a herb garden beside the sinking patio, where we would enjoy the good life and 'grow our own'.

Out of the window, then, flew my plans for a fantastic tree house, a garage for my classic Jag (still only a figment of my dreams and wishes) and a Victorian bathing hut. It looked like the herb garden was going to happen. It was to be the perfect new outlook from the kitchen window and it would be situated on the eight-foot squared site of wilderness and weeds, where the long grass came up to my knees. I did not envy the man who was going to transform this area, but I did provisionally hire a rotavator to help clear the ground for him. I was advised that the rotavator was the wrong machine for the job, as it would just turn the soil and the turf and the grass and the rocks and there would be no real clearance on which to build the herb garden, so I cancelled the machine and hired a Herculean labourer to remove the entire surface, of which there was a lot, and carry it off to the woods manually. I think this patch must have been a rockery at one time, because we managed to accumulate enough boulders to build a cairn. Our excavations also unearthed a workman's boot (c.1958), an old tool from the Iron Age, a full tin of Germolene that still smelled of Dalkeith Thistle's dressing rooms, a beautifully moulded, unbroken bottle of Sloan's Liniment, and a Victorian jar top whose decorative label reads 'J.B. Mitchell – Pharmacist – successor to D. McLaren – 58 South Clerk Street, Edinburgh'.

After eight hours' hard labour the ground was flat and ready to work on. Ricky, the gardener with the vouchers, came from Newtongrange, next door to my Auntie Catherine, and set about planning out a much more inviting visual aspect than had previously been there. Part of the peripheral plan involved building an arched trellis which led to the 'back' end of the garden that's nearly in the woods, and climbers would hopefully climb all over it one day. As he was nailing this thing together, Ricky disturbed a wasps' nest by accident and the evil monsters went after him. The only protection he had was the hammer he held in his hand, and in a state of panic he tried to hammer the wasps away from his face. Unfortunately, very few wasps were hit but Ricky did manage to bash his own head, and the day kind of deteriorated from there. The wasps dispersed when Ricky ran at great speed to the other end of the garden until at last he was able to stop and assess the damage. He jumped in his car and drove himself to casualty, where he was treated for a few stings and a bleeding head.

At the risk of sounding like the softest character in *Little Red Riding Hood*, all this time I had been at the garden centre to buy some herbs for the garden. When I returned, I noticed Ricky's car had gone and a note was attached to the top of the wall, held down by a half brick, written in blood, telling me there had been an accident. I looked around at the aftermath of the disaster trying to work out what could have happened: big yeti footprints in the smoothed out soil, a half-finished trellis, and a trail of blood from the hammer (which looked like a murder weapon) to where the car had been parked. I wasn't sure what to think, but I was kept in suspense until Ricky returned.

When Ricky came back, somewhat recovered from his ordeal, he pointed out the remains of the nest, right beside the half-erected trellis arch. There was still sinister activity in the tree and I was amazed that I hadn't spotted the nest before. It seemed so obvious now, but for the rest of the day Ricky kept well away from it as he planted herbs. I phoned the Council and a man from the Environmental Health Department came up dressed in a NASA space suit, armed with some heavy duty fly spray and an extended artificial arm. He fumigated the derelict nest and exterminated the remaining wasps, who were still clearly pissed off. He then removed any trace of wasp from the tree and Ricky was able to get his job finished, the result being one fine herb garden whose produce has enhanced many a bowl of Heinz tomato soup.

The financial outlay when doing up a house is never ending. If it's not roofing repairs, rebuilding chimneys or landscaping, it's the more mundane necessities like servicing the central heating system to comply with the terms of the guarantee agreement, or the regular protective painting of fences and gates, or the upgrading of the burglar alarm, or tree surgery or draught-proofing, or simply the redecoration of another room because its time has come round again after five years. So why bother?

Thinking about it, Arlene and I like to have a nice place to go after a day's work or on a day off. It's our great escape. It's a fun place for Caitlin and her friends and cousins to grow up and play. It's a pleasant environment for entertaining our own friends. It's away from it all, yet so close to the buzz of the city. It's where *I* would go for my holidays. It's something different.

I have lived in student squalour, in other people's pockets; crashed out on couches and floors rancid with curry stains, damp with beer; slept in Transit vans. I have lived alone in a two berth caravan in sub-zero temperatures with fungus creeping up the wall, no water supply, heat or electricity, and no money to justify paying some hood an extortionate rent. I have experienced lack of privacy with other people's rotting dishes piled up in the sink for weeks, their choice of music and television; their weird acquaintances. I have put up with the smell of boiled cabbage in my flat for three terms, tolerating cultural differences between Scotland and China beyond the call of duty. I have waded through other people's dog shit to hang a towel out to dry in my *shared* garden, and I have been kept awake by babies, budgies and dogs at close range through paper thin walls. I have heard Queen's Bohemian Rhapsody nineteen million times in succession at full volume from next door with three teenage girls singing along, and experienced Mozart's Rondo à la Turk on a honky-tonk piano the same number of times. I still have flashbacks to the sound of Woodburn Miners' Club resident band playing 'Yellow River' by Christie for a slosh *every* night throughout the early seventies as I lay in my bed, and I have even slept in the back of a Marina Estate on a regular basis while the snow lay deep and crisp and even outside, the alternative accommodation at the time being so bad that it was my second choice. Some of this I have actually enjoyed, as it has come with the lifestyle I led at the time, and in fairness, I'm sure I must have thrown back *some* of these unpleasant habits to my own neighbours and flatmates over the years (loud music, dodgy piano playing, weird friends) but now it's payback time. At whatever the cost, I am happy to occupy this relatively small, comfortable space on the planet where I can be loud or quiet, sensible or eccentric, keep dogs or not. I can go to bed early and sleep in peace, or I can have a heavy metal band jamming in the living room at four o'clock in the morning if I like.

At the time of writing this, we have already lived in The Old Windmill longer than Elizabeth Adair's entire stay here, and you come to realise, or at least to expect, after years of hard work, that the value of the property will have increased. Anyone who ever buys a house hopes, if not expects this to be the case, and most of the time houses do increase in value over the years. I remember how difficult it was for me to raise enough money to qualify for a mortgage on my first tiny flat in Aberdeen. It cost £10,000 then – a fortune – and thanks to my parents giving me a loan for a deposit, I managed to buy it. Three years later I sold it for £12,000 and I poured the massive profit of two grand, along with Arlene's life savings, into the purchase of our first flat together – the one in Blenheim Place. As you know, a miraculous profit of thirty-five grand was made on that in just three years, and so it goes on. Inflation seems to mark time in its own way by making yesteryear's prices sound ludicrous. It's a bit like when my Granny used to tell me how she could get into the cinema for jam jars. Try that trick at the UCI nowadays and the spotty youth behind the counter wouldn't even get the joke. Comparatively, it wasn't so long ago that brand new luxury cars cost less than a thousand pounds on the road; go back just a little bit further and a three bedroom house cost the same. When I was young, a bag of chips was four pence (that's *post* decimalisation) and now, not that much later, chips are almost a pound a bag. So relate all that to house buying, and the price tag on The Old Windmill in 1962 will amaze you.

Elizabeth Adair's mother lived in the Cotswolds. Her health was failing and Elizabeth had to go and look after her, and so she was faced with the frustration of giving up the windmill so very soon after all the effort that gone into making the building habitable; all that determination, her passion, her vision. When the time came for her to sell The Old Windmill it went on the market for £8,000. I have not missed out a nought: £8,000 – for a restored, converted windmill with all mod cons in nearly an acre of splendid garden in a delightful village near Edinburgh! What is more unbelievable is that only one insulting offer of a mere £4,000 was submitted at the time (that's *less* than the value of my fourteen-year-old car today) and an ice-cream vendor suggested a furnished lease for a year, but needless to say he was booed off too. Stop me and buy one. The Old Windmill was re-advertised three months later and the Chalmers family had the good sense to snap it up on returning home from abroad.

THE OLD WINDMILL

CHESTERHILL, FORD, MIDLOTHIAN, SCOTLAND

Ten miles south of Edinburgh on bus route.

A house of great distinction featuring the restoration of one of the few remaining windmills of the 18th-century in Scotland, allied to a modern wing, architect designed, in keeping with the character of the original tower, which is on the list of Buildings of Architectural and Historical Interest in Scotland.

Tempting, but no takers: the schedule of 1962.

Only one year before, Elizabeth's house-warming party was in full swing, attended by friends and relations, representatives from the BBC, Historic Buildings, Ancient Monuments, and Homes and Gardens. In fact H&G were first to do a complimentary feature on the house and many others followed suit. The restoration and conversion process was also the subject of a BBC television documentary in the very early sixties, but this no longer exists – believe me, I have scoured the archives.

It must have seemed like no time at all before the removal men were back, packing up the endless books and pictures and china. At six o'clock in the morning, one day in the middle of September 1962, Elizabeth Adair left The Old Windmill and drove south in her packed-to-the-gunnels Mini van, somewhat out of pocket, after a remarkable adventure.

It's incredible to think how desirable and valuable converted old buildings are today. A friend of ours, a professional architect, recently evaluated The Old Windmill in its present condition at almost double what we paid for it only five years ago, and in turn Elizabeth Adair probably never dreamed of seeing us, or anyone else, buy it for *twenty* times its 1962 value. Somewhere down the line, I'm sure, even the figures we are talking about today will seem like sweetie money to someone else.

As early as our first summer at the windmill, Dad and I had ideas for another extension. I remember saying to him that these were 'long term' plans and they probably wouldn't materialise for another five years. Five years seemed like a lifetime away then, and I cannot believe how quickly they evaporated, nor the changes they have seen. I could never have dreamed that Dad wouldn't be here in five years' time, or that within only a year of his passing his first two grandchildren would be born – Caitlin in March, then Jacqui and Gerry's wee boy Michael in November. Lousie, Michael's sister, arrived not long after and then at Christmas time came our wee Robbie, a brother for Caitlin.

Occasionally I still stand on the site of our make-believe extension, contemplating our design, looking at the practicality, and then shuddering at the cost. An extension would ideally be the next phase, but realistically, lack of money is likely to prevent the work being done within the next five years, probably even in the next twenty. It doesn't hurt to dream though. If a windfall was to come our way, perhaps this is how The Old Windmill would look in the new millennium:

You walk into the recess from the hall under the main stairs where the telephone table is, and the wall has been knocked through to form an entrance to a new wing. To your right there is a 'back' door framed on the outside by a small traditional timber porch, a New England-style mud room where muddy wellies, umbrellas and coats are kept. A flagstone path runs all the way from the door past the back of the house and right round the tower, bringing you out at the drive at the front. Back to the entrance to the new wing, and the small vestibule you walk into has an interior door straight ahead with a staircase leading off it. The door takes you into a large rectangular room with natural wooden floors covered with some colourful rugs. In the middle of the wall facing you there's a small sandstone chimney with a wood-burning stove at the foot of it. On your right, a bed folds down out of the wall and remains inconspicuous when you don't need it (I saw one of these in a James Bond film years ago) and the bed is flanked by two slim windows either side which offer a view of the tower with its backdrop behind and, of course, the sunrise in the morning. On the left wall, huge patio doors slide open, leading down three old stone steps into the most private bit of the garden, where the herbs are, then on to the patio, with a pleasant view of the colourful rockery, now mature and sprawling, and Caitlin's old Wendy house. The room is simply furnished with a strong Scandinavian feel, but it's functional and multi-purpose, bright and cosy. Two large comfy chairs swivel to face the stove at night as their occupants read in utter peace, or if you prefer, the chairs can turn to face outside, where you can watch the playful squirrels or the timid deer up close, unaware that they are being spied upon. If you double back on your left to the bottom left-hand corner of the room you enter a small en-suite shower room housed under the new stairs (which we'll come to) and again this is bright and clean with one window that looks out across to the woods. Back to the point of entry once more, and from here, immediately to your left, a simple white wooden construction takes you up a flight of stairs that runs adjacent to the existing dining-room wall, and at the top of the stairs you are met by another small window which happens to be directly above the window in the shower room. The light floods in and down the staircase, illuminating the whole of the new vestibule, and you can see right across the garden from the top step. From here you take a right through the door into this upstairs room. It's a small but adequate bedroom-cum-playroom for Robbie with all the spoils a child would want in it: drumkit, Scalextric set, books, computer, toys. The room is lit by two velux windows in the roof giving a perfect view of the sky above, slightly overhung by the tall trees, and at night (looking up from a snug bed) the big old moon and a hypnotic galaxy of stars shine and wink at each other across the universe.

Then you suddenly think of all the plumbing, the wiring, the heating, the planning permission, lawyers and builders. Things go wrong. The budget goes out of control! The bank manager enters your daydream, uninvited. You think of a big number and you double it then multiply it by six, and then you realise the price of your brilliant idea is exorbitant and, certainly for the foreseeable future, unaffordable.

Meanwhile, back on Planet Earth, I sat on the wicker chair under the main stairs in the arched alcove where my Granny's Singer sewing machine serves as a telephone table, and everything was back to normal. (I was jolted from my stargazing by two rings of the telephone, but I decided to let the answering machine take the message.) These were the plans Dad and I had enthused about as we paced out the rough measurements along the back wall that summer. Part of me has always thought the extension idea was pie in the sky but not so long ago, so too was the idea of living in a windmill at all.

Can you fall in love with a building? This was the question Elizabeth Adair asked herself when she stumbled across The Old Windmill in 1958. She wrote the question in her book. I don't really know what her answer might have been and I never thought to ask her when we met.

To me, the house has a bewitching quality and a shadowy past. Its setting is inspirational, charming and magical, and to say it is unusual is an understatement. Undoubtedly, there is nowhere else I would rather live. Roll out the clichés and home is where the heart is, home sweet home, sure. And although we've settled here, it doesn't necessarily mean our travels and adventures are over but I have more than a feeling that this is where we were meant to be, and maybe someday we'll be the old folks that live on the hill.

Chesterhill Mill's sails stopped turning *long*, long ago, and time marches on so quickly; and who knows where we'll be a few years down the line; but if the winds of change turn out to be a breeze, with a little luck The Old Windmill will be our family home for many years to come.